The Path to Pivot

The Playbook for
Founders Who Want to
Reboot Their Startup

Jason Shen

ISBN (paperback) 979-8-9906104-0-8
ISBN (ebook-ePub) 979-8-9906104-1-5

ISBN (ebook-KDP) 979-8-9906104-2-2
ISBN (paperback-KDP) 979-8-9906104-3-9

ISBN (hardcover) 979-8-9906104-4-6

Refactor Labs
367 St Marks Ave
Unit #795
Brooklyn, NY 11238
refactorlabs.xyz

To every founder who dared to dream.
Go forth and write a new destiny.

"Not everything that is faced can be changed but nothing can be changed until it is faced."

— *James Baldwin*

"You never change things by fighting the existing reality. To change something, build a new model that makes the existing model obsolete."

— *Buckminster Fuller*

PIVOT TOOLKIT

APPENDIX

INTRODUCTION

My tale of two startup pivots—one failed, one successful—and how this book will help you reboot your startup.

The highest and lowest moments of my first startup venture were barely a month apart.

In August 2012, my team and I were part of a Vanity Fair cover article photoshoot alongside the Dropbox founders and the partners of Y Combinator (YC) with the headline "Who Wants to Be a Billionaire?"

A year earlier we had launched our company Ridejoy out of the Silicon Valley accelerator with a long-distance ridesharing service. Back then Lyft didn't exist yet and Uber was an app for getting limos in San Francisco.

A journalist was embedded in our YC batch of summer 2011 to write a book about the world's most successful accelerator. They featured Ridejoy's story on the very first page. Vanity Fair was running the first chapter of this book as an excerpt hence why my co-founders Kalvin Wang, Randy Pang, and I got to pose next to Silicon Valley royalty as the rag tag team of young upstarts.

Beyond this ego-stoking experience, we had also just launched our iOS app after 7 months of development and it had quickly been featured by App Store editors as an eco-friendly way to travel. We were soaring high and the future looked bright.

Just a few weeks later, we were crashing back down to earth. Our only effective growth channel, Craigslist, had threatened legal action to keep us from siphoning users off their platform, and we needed to make some big changes.

My cofounders and I tried to explore adjacent opportunities in the travel space. When that exploration went nowhere, we were forced to lay off the team to preserve cash and retrench. We spent eight months jumping from idea to idea, unable to agree on a direction to start building.

We didn't run out of money—we ran out of resilience.

I remember huddling over the phone with my cofounders as one of our investors, Garry Tan, now President of Y Combinator (gulp) chewed us out for wanting to give up. "Do you know how many founders would kill to be in the position you're in?"

We should have been able to pull off a pivot, but we failed. If I had known what I know now, I wonder if things would be different.

⁂

This book is written for founders and CEOs of venture-backed startups contemplating—or in the middle of—a major shift in their business.

If that's not you, that's cool. Maybe you're an indie hacker, corporate executive, or angel investor. There is still something here for you, but I would recommend you to tweak the advice as needed.

I've taken my own advice and written for a specific niche: the early / mid stage startup funded by VC dollars. My target reader is the person in the arena[1], face marred by dust, sweat, and blood, facing a tough decision. This book is for you.

When conviction fades

Dear founder,

You've probably been in business for a few years, having built a modest level of traction. A good number of consumers or companies use your product—maybe you're even generating meaningful revenue from these customers.

You've raised a few million in venture funding, made a name for yourself in this space. On the outside, you look like a successful

---

[1] In the sense of Teddy, not Chamath
https://knowyourmeme.com/memes/in-the-arena

company. Your friends and family have been following your journey with enthusiasm.

Despite all your external success, you've got a problem. A nagging doubt in the back of your head, speaking in a whisper or perhaps even a shout, that you're going down the wrong path.

You've missed major milestones you've set for the team and the gap seems to be growing every day. You're watching your bank account drop without knowing how you're going to raise more money.

Maybe you're having trouble sleeping. You've got a perpetual tightness in your chest. Headaches that knock you out for hours. A distinct lack of appetite or enthusiasm for even your favorite things.

Conviction is one of the most important and powerful assets a founder can possess. It's what inspires your team, recruits new customers and employees, wins deals, and lands new investment. But these days, your flame is getting weaker and dimmer each day. Intrusive thoughts are popping into your head:

- Do I really have what it takes to succeed as a founder?

- Would my investors freak out if I tried to radically change the business? What about my team?

- Why the hell did I quit my job to chase this dream?

Whether you came across this book via a desperate search for an answer out of this predicament or because someone you trust recommended it to you, rest assured: you're in good hands.

Over a couple of short, actionable chapters, I'm going to share a practical framework for rebooting your business, without burning all your bridges or imploding your business in the process. Because I've done it.

The tale of two pivots

I wrote this book as a resource I wish I had when I was building my companies. Everything about a startup is hard but pivots are particularly uncertain because there's such a wide range of opinions and anecdotes about how to execute one.

I attempted a pivot in my first startup and failed, ultimately shutting down due to our inability to commit to a new direction.

My second startup also hit a point where a pivot was necessary. This time, I avoided the mistakes we made the first time and successfully navigated through the pivot, raised new funding, and eventually landed a modest acquisition by Facebook.

Pivot 1: Failing out of fear

Despite warnings from Paul Graham that ridesharing was a bad business because the market would be limited to people who were extremely cost-conscious, my cofounders and I launched Ridejoy at Y Combinator Summer 2011 Demo Day. We were thrilled to raise $1.3 million for our long-distance ridesharing platform, which was about average for YC back then. Though some of our peers like Codecademy, Parse, and (Rap)Genius raised far more).

We launched as a carpool service for Burning Man and got 1,600 users without spending a dollar on advertising. Things were looking great. But there was a problem — while we easily acquired users through Craigslist, we struggled to gain traction from other channels like events and partnerships, and couldn't inspire word of mouth. Our Ridejoy iOS app also failed to drive growth despite an initial press spike and getting featured by the App Store editors. Then, Craigslist sent a cease and desist for letting users post back to their platform—our growth hack was toast.

With no clear path to Series A, we tried pivoting but the process was chaotic and our employees were clearly rattled and unprepared for such a sudden shift. We let our team go, gave up our office, and retreated to our shared apartment to find a new direction. We had built a warm, close-knit culture with home cooked meals at the office and watching our employees cry on our last day as a team was one of the worst feelings I've ever experienced.

We struggled to come up with a new direction. I cringe at our ideas now: a social media platform for sharing thoughtful quotes, a tool for pooling donations to charity, a laundry delivery service.

As privileged twenty-somethings with little business or life experience, finding "big problems" that we were uniquely qualified to solve felt next to impossible. Garry Tan saw it differently. "Just call up some YC companies, ask them what problems they have, and go solve those!"

We were naive and thought that we could dream up an idea out of our living rooms. We were wrong.

Without a structured framework, a deadline, or any real accountability, we were stuck. We ultimately returned nearly six hundred thousand dollars back to our seed investors.

Result: *Failed pivot, business shut down*

Pivot 2: Succeeding through scrappy execution

Nearly five years after the Ridejoy debacle I founded a new company called Headlight with Wayne Gerard, an engineer I worked with at Etsy. We wanted to reinvent the hiring process by making it easier for companies to assess a candidate's technical skills via a timed take-home assignment. After raising a small pre-seed round, we landed a handful of customers for our initial software product.

Following market demand over the next year, we found ourselves operating more-or-less a technical recruiting agency rather than a tech platform. My cofounder and I felt like we were no longer building a venture-scale business and wanted to shift gears.

When our lone in-office employee went to a conference in Europe, we took a week to map out new directions we could take our business that we presented to a small group of trusted advisors. One area that excited us was gaming—my cofounder Wayne had been a Diamond ranked Starcraft 2 player and worked at Bloomberg Sports, an analytics platform used by professional sports teams. I was more of a casual gamer but loved the growing competitive scene in professional and collegiate esports, which reminded me of my own experience as an NCAA athlete.

Throwing away our brand, product, and market position was painful, but we let our conviction guide us. Despite our lack of experience or relevant product offering, we pursued a hard pivot in the gaming and esports space.

We renamed the company Midgame, raised new capital from Techstars, Amazon, and Betaworks, recruited a new team, and built several analytics and proto-AI powered products that were used by collegiate and professional esports teams as well as consumer gamers. It was thrilling to build in a new space, going from B2B to B2C and avoid the stallout that Ridejoy suffered from.

That said, we struggled in other ways: raising a new round with a crowded cap table was tough and our lack of experience in consumer growth marketing held us back. In the end, we weren't able to hit product-market-fit with any of our product offerings.

After the pandemic hit, investor funding dried up right as we needed to raise more capital. It was a tough break, but we solicited offers from several companies before taking an acquisition offer from Facebook to bring our team into the social media giant.

Result: *Successful pivot, business was acquired*

Note: There's more to my pivot stories but rather than get too into the weeds now, I'll share the details as they become relevant in the pages to come.

From gymnast to founder to coach

I grew up in a home of coaches & educators: my dad worked in education policy and founded a local Chinese language school, while my mom was a gymnastics coach and high school teacher.

We didn't have much growing up but I'll always appreciate being one of the first kids in my neighborhood on the internet—dialing through a 14.4k modem on a Packard Bell desktop in the mid 90's. I got little exposure to business through them, but lots of exposure to developing human potential.

As a competitive gymnast, I was transformed by coaches who challenged me and nurtured my latent talent—it was my gymnastics prowess that got me a scholarship to compete for the gymnastics team at Stanford. I left 5 years later with a masters degree, an NCAA championship ring, and a newfound passion for entrepreneurship.

After Ridejoy I worked in the Obama administration as an innovation fellow at the Smithsonian, backpacked in Peru and relocated to NYC to improve my dating life. After two years in product at Etsy, I was let go in the second wave of mass layoffs by the online retailer. While disappointing, this kick out the door was what I needed to get serious about launching Headlight.

I was already working with an executive coach (shoutout Hamilton Chan!) who helped me stay grounded through the entire Headlight/Midgame journey, especially in hard periods like the pivot and the acquisition.

At Facebook / Meta, I worked across a number of product teams, starting with knowledge and documentation tools meant to drive productivity among internal tech teams. Eventually I made the switch to the consumer side, building experiences on Facebook Groups that reached hundreds of millions of users every day. The work was invigorating but day to day work of dealing with reorgs, new project shuffles, and changing leadership directives was draining. I missed the freedom of being my own man.

A year into Meta, I started my own coaching practice on the side working with founders, engineers, UX researchers, even other coaches. I found myself more fired up about working with my clients than any of my projects at work, dare I say even more than being a founder myself.

I wanted to try something new and connect to a mission that really mattered to me. In June 2023, I resigned from the company to coach full-time and write the first version of this book.

There are a lot of anecdotes about startup pivots, but nothing that comprehensively helped founders go from pre-pivot to post-pivot. I've always admired books like The Lean Startup, The Hard Thing About Hard Things, and The Mom Test. Books that offered tangible lessons for startup founders based on the author's experience in the trenches. Resources that have stood the test of time and are passed from founder to founder.

After looking for such a book around startup pivots, I found nothing. So I decided to write one that might someday join that pantheon of books. Only you as the reader can judge whether I was successful in my goal.

Working with me

Today I engage with 15-20 founders, senior leaders, and outlier achievers in New York, Silicon Valley, and around the world. We work together to unlock their full potential, lead through difficult transitions (e.g. pivots, exits) and make their dent in the universe.

If you are a high achieving founder and think executive coaching could make a difference for your business, you can learn more at **jasonshen.com/coaching** or email me at **jason@jasonshen.com**.

Pivots are smart, not shameful

Some founders have it in their heads that pivoting means they were wrong, that they made a mistake with their first idea and that they're a failure somehow. Nothing could be farther from the truth.

Here's a quote from Kevin Systrom, who pulled off one of the most successful pivots of all time, Instagram:[2]

> *"Far too many people, because of ego, stick with ideas far too long and it ends up really poorly. And the entrepreneurs that I've seen do very well are the ones that are equipped and engaged in the change from something that's not working to something that is."*

And again from seed investor and Y Combinator founder Jessica Livingston, who interviewed dozens of founders for her seminal classic Startups At Work:

> *"Founders need to be adaptable. Not only because it takes a certain level of mental flexibility to understand what users want, but because the plan will probably change. People think startups grow out of some brilliant initial idea like a plant from a seed. But almost all the founders I interviewed changed their ideas as they developed them."*

So not only are pivots completely normal and not a sign of failure, but in fact the ability to adapt your plan in the face of new information is crucial to success as a founder.

Plotting the path to pivot

The main idea of this book is simple:

If your startup isn't working, don't wait until you've got just a few months of runway left to make a desperate gamble on a new

---

[2] Some estimates place Instagram's value as a standalone business at $47B as of 2022 https://tim.blog/2019/04/25/kevin-systrom/

direction. Instead, think about pivoting sooner, and slower so you can make the best decision and bring everyone along with you.

This book is broken into three major sections:

1. **Pivot Blueprint:** This is the heart of the book, focusing on the strategy and mindset you need to make the leap from a stagnant business to one with a path to exponential growth. These six chapters cover the Align-Explore-Commit framework, the idea of a pivot pilot, how to make the final decision choosing to pivot, and explaining the decision to customers, team members, and investors.

If you only get through this section of the book, you will already be far more equipped to make the leap.

2. **Pivot Toolkit:** In this section, you will find practical strategies and actionable steps to execute your pivot with confidence and impact. We cover getting stakeholder buy-in, managing the finances, navigating the idea maze, and staying resilient.

Feel free to skip around, read out of order, and focus on the chapters that are most relevant to your situation.

3. **Appendix:** Here you will find some supplemental information, including some calculations about Venture Math and why you need to swing for such a massive result, and a set of detailed case studies of successful pivots.

I've tried to make each chapter concise and self-contained while including stories of successful pivots by companies you've heard of, and ones you haven't.

Countless entrepreneurs have been right where you are: unsure of your direction but not ready to give up. It's not too late, but there's also no time to waste. Let's get started.

Recap of Introduction

What happens when you lose conviction as a founder

- Conviction is your biggest asset and it's scary to admit when you've lost it

- I've lost conviction twice. (i) Ridejoy: Failed out of fear (ii) Headlight/Midgame: Success with scrappy execution

- If you've lost conviction, accept reality and get to work on a new plan

Plotting your path to pivot

- Don't wait until you have a few months of runway to make a desperate gamble. Pivot sooner to give yourself time

- Give your pivot structure with the "Pivot Pilot" approach

- A pivot includes the existential (finding your next big idea) and logistical (winding down your business, rebranding, etc.) We'll go through it all

The Path to Pivot Framework

PRE-PIVOT:
- **Lost** conviction
- **Slowed** growth
- **Limited** total opportunity
- **Shaken** in market changes

POST-PIVOT:
- **Renewed** conviction
- **Accelerating** growth
- **Expanded** total opportunity
- **Capitalizing** on market changes

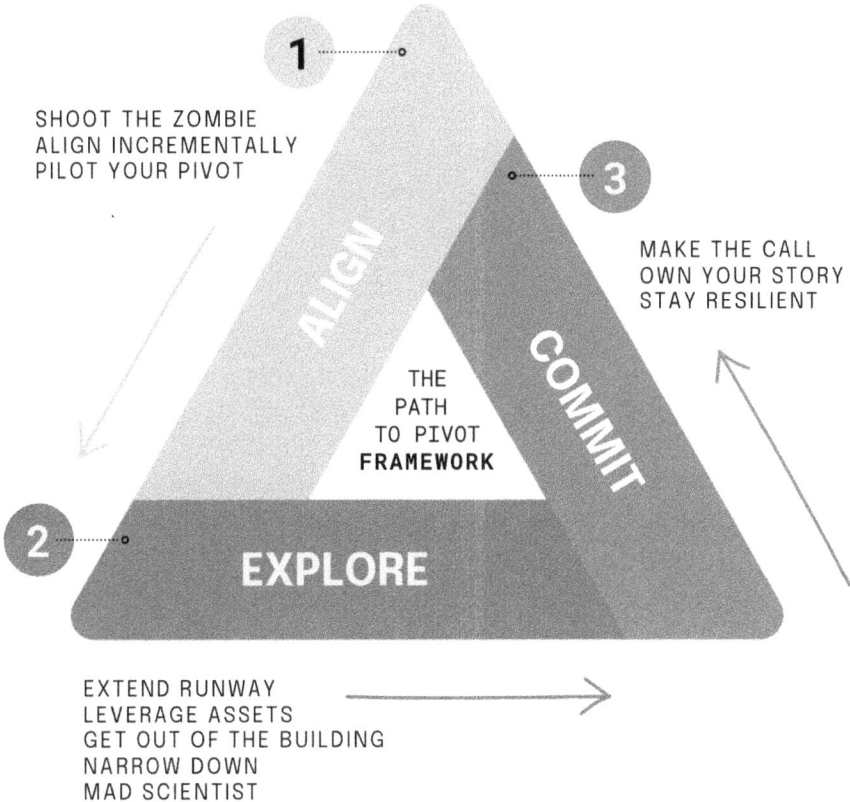

1

SHOOT THE ZOMBIE
ALIGN INCREMENTALLY
PILOT YOUR PIVOT

3

MAKE THE CALL
OWN YOUR STORY
STAY RESILIENT

ALIGN

COMMIT

THE
PATH
TO PIVOT
FRAMEWORK

2

EXPLORE

EXTEND RUNWAY
LEVERAGE ASSETS
GET OUT OF THE BUILDING
NARROW DOWN
MAD SCIENTIST

Download a high-res PDF of this framework, along with videos, templates, worksheets and other free pivot resources at

pathtopivot.com/resources

"Everyone thinks of changing the world, but no one thinks of changing himself." — *Leo Tolstoy*

Section I

PIVOT BLUEPRINT

In this section, we lay the strategic groundwork for a successful pivot—the mindset, the time-bound pilot approach, the decision making process, and the art of communicating your change.

If you only get through this section of the book, you will already be far more equipped to make the leap

CHAPTER 1

GRIT, QUIT, OR PIVOT

We are taught to push through every obstacle, but sometimes the right move is to play a different game—don't let biases cloud your judgment

Airbnb CEO Brian Chesky looked like a shell of a man.

Our cohort of YC founders were gathered for our Tuesday dinners, where we'd eat home cooked meals (a lot of chili) and hear off the record stories from successful founders and investors.

YC partner Paul Graham and Chesky were seated across from each other on stools, at the all-orange Y Combinator headquarters back when they were in Mountain View. PG had a jovial smile, with his trademark sweater and cargo shorts, while Chesky wore a black t-shirt and jeans, heavy circles under his eyes.

At the time Airbnb was already a startup darling, having booked 2M nights on their platform, raised several major rounds of funding and enjoyed largely positive press coverage about their form of short-term housing that was a fresh alternative to hotels, with a local feel, where real relationships were being formed. But earlier that summer, two high profile incidents of hosts having had their homes ransacked and damaged had battered the company's reputation as a safe way to earn money and meet friendly travelers[3].

While I can't share certain details from their conversation, the aspects that stand out even now are how long the road to success had been for the founders back in the day[4] but also even then, with their established status as a breakout startup on a clear trajectory towards a billion dollar company and beyond.

After the two nightmare host incidents took place, Chesky realized this was a crucial make-or-break moment for the company. He described pulling the company into "lockdown mode" to tackle the problem. Several weeks later the team emerged with a $50,000 host guarantee. Airbnb knew this was still just a stop-gap and to further ensure host confidence, they announced a $1M host guarantee the next spring.

Chesky and his cofounders at Airbnb could have quit at many points during the journey, but they stuck it out, to great success. Airbnb is one the most valuable startups in the entire YC portfolio of 4,000+ companies.

In this chapter, I will try to convince you to pivot your company.

But to be fair, we should look at your other alternatives and make sure it is in fact the right choice.

---

[3] https://techcrunch.com/2011/07/31/another-airbnb-victim-tells-his-story-there-were-meth-pipes-everywhere/

[4] Airbnb was legendary for having raised $40,000 selling branded cereal as a desperate attempt to stay afloat while trying to raise their first institutional round of funding. Apparently they had at that time a "binder full of credit cards" that they were using to keep the company alive—obviously not something anyone can advocate for.

Grit: First off, sticking it out is always on the table. Airbnb could have easily folded at various points in their early life as a company. I believe there are founders out there who are constantly changing ideas and need to hear that they actually need to dig deeper with their current approach.

> *"A big failure I've seen is a founder quitting right after launch and not grinding out the execution. You need at least 4-6 months on an idea before you can definitively say it won't work."*
>
> *—Bilal Mahmood, founder of ClearBrain (acquired by Amplitude)*

Quit: We should consider the path of quitting. As terrible as it might sound, there are times when giving up on your company entirely (or at least leaving it in the hands of your remaining cofounders / team) might be the best choice.

Pivot: If you made it this far, you've already thought pretty hard about making a major change to your startup's strategy or direction. We'll spend most of the chapter talking about this option, but again, it's not the only one.

So let's talk about it.

Option 1: Grit

Startups are hard and sticking out through the grind, the trough of sorrow, is important. You've probably seen this chart by Paul Graham[5] and you're struggling in that long bottom part that feels like it'll never end.

There are many stories to choose from—let's take Stripe. It took the fintech company two years from their first line of code to public launch as they had to secure a number of partnerships with financial

---

[5] This cleaned up version is courtesy of a16 partner Andrew Chen
https://andrewchen.com/after-the-techcrunch-bump-life-in-the-trough-of-sorrow/

institutions. Along the way they installed their API manually to customers and even after that public launch, it was not obvious that this young company could ever expand beyond a small niche of developer-first internet companies to dominate the payment processing landscape. But if the Collison brothers had given up along the way, Stripe wouldn't be one of the world's most highly valued startups today.

So should you just keep going? Here are some indicators that gritting it out is the best decision:

1. **You haven't launched or are still on V1.** What are you doing my friend? Before you can say your product or idea isn't working, you gotta get it in front of real people. The number one question Paul Graham (and now the other partners) would ask during Y Combinator to startups that have not launched yet is "When are you launching?" That's what you'll hear on repeat until it happens.

   Unless you've realized that your product is completely infeasible given your team and funding, you should first ship your V1. Cut scope dramatically if you have to, but ship and iterate at least a couple rounds over the next 4 to 6 months before considering any other moves.

2. **You have only tried 1-2 sales or marketing channels.** Getting products out to market is hard, and selling is often uncomfortable for technical or product-minded founders. Far more companies fail because they weren't able to generate demand than because their product just didn't work.

   When building my second company, Headlight, we used content marketing, speaking events, panels, and cold outreach to seek out customers for our tech candidate evaluation product. We weren't afraid to get on the phone, even meet clients at their offices and we iterated on our offering multiple times. By the time we decided to pivot, it was clear that we had exhausted most of our options (though looking back I think we could have done more here).

3. **You haven't tested the most risky assumption.** Later in this book we'll talk about how you need to think like a scientist. Many founders are familiar with data-led product development but it's worth reminding ourselves of what we meant to test.

   Astro Teller, who leads moonshot projects at Alphabet's X division, talks about how you need to "tackle the monkey first". This refers to the idea that if you are trying to train a monkey to get on a pedestal and recite Shakespeare, you should spend 100% of your time training the monkey and 0% of the time building a pedestal. Because that's the hardest part of the idea[6].

   When we were building Ridejoy, we had several common excuses for our low bookings: people couldn't find the right ride, they didn't trust the driver, etc. One of our advisors had an idea: build a fake version of the app with artificially arranged perfect rides to see if people would click the "Arrange Ride" button more. It was a compelling suggestion but we chickened out. We were afraid to find out what would have happened if we did it and people didn't click. We weren't ready to face reality. Too bad. It would have quickly given us reason to keep going, or a clear signal to change course.

---

[6] Not dissimilar from Mark Twain's advice that if you've gotta eat a live frog today, do it first thing in the morning. Twain probably would have made a great founder.

4. **You are fighting with your cofounders.** Like a couple hoping that having a baby might fix their relationship, some founders see pivoting as the solution to their constant conflict. It's not. The excitement of the new idea might relieve some tension initially, but the stress will return, and with it, the conflict.

   Use this opportunity to step back from the day to day work and address the underlying issues—usually involving some sense of unfairness or imbalance between your roles, responsibilities, and decision making authority. Maybe pivoting is still in the cards, but don't use it as a way to avoid a hard conversation.

5. **You are exhausted and haven't taken a break in months.** Burnout is real, and it can cloud your judgment. If you're feeling drained, it's tough to know whether the challenges you're facing are insurmountable or just seem that way because you're running on fumes. Take some time off to recharge before you make any drastic decisions about the future of your startup. A clear mind can do wonders. It could reveal solutions you hadn't previously considered or give you the renewed energy to tackle problems head-on.

Option 2: Quit

Quitting is often seen as the last resort, but sometimes it's the most sensible option. Founders often carry a ride-or-die mentality. Whether it's a chip on their shoulder from former classmates, colleagues, or family members, a desire to prove something to themselves, or fear of being seen as a failure, the powerful trait of commitment and dedication can backfire and cause founders to stick to things too long.

Former professional poker player Annie Duke observes in her marvelous book Quit that it's easy to miss the clues that quitting is the best choice because "in practice things generally won't look particularly grim". It's our aversion to quitting that keeps us from seeing the writing on the wall.

1. **Team Breakdown:** There's a difference between team conflict and irreparable disagreements. Cofounder conflict is the number one killer of early stage startups according to YC President Garry Tan.[7] Tan described having a fundamentally broken relationship with the cofounder of his first startup Posterous—struggling to eat, sleep, or have a calm conversation about the future of the firm. When trust is eroded to the point where you question each other's decisions or even integrity, it may not make sense to continue. Sometimes you can fix this by removing one of the founders. In Tan's case, he left the firm with his cofounder, but if you can't get to that alignment, often the only path forward is shutting down.

2. **Personal Toll:** If the startup is taking an irreparable toll on your health or familial / marital relationships, it might be time to consider whether it's worth continuing. I once saw a founder share that they had put off a big toothache for several weeks and when they finally went in to see the dentist, they had to complete emergency surgery for a very bad infection that left him in the hospital for weeks. The guy narrowly dodged a bullet but this mistake could have left permanent damage or worse. If the demands of your company are truly this severe, you need to get out.

3. **Insurmountable Obstacles:** Sometimes you hit a wall that's not just high, it's seemingly insurmountable. I'm talking about regulatory hurdles that would take years and capital you don't have to clear. FDA approval for pharmaceutical

---

[7] https://techcrunch.com/2017/02/18/co-founder-conflict/

drugs are notoriously long and if you fail a clinical trial, you might not have enough to complete another one.

Or maybe a dominant player's new release makes your product redundant, and they have overwhelming market reach. Many a company has been killed by a big update by Apple or Google, and even Slack was eventually beaten by Microsoft Teams.

4. **Less than 3 months of runway.** Look, while there are times when you should fight until the bitter end, if you are pretty sure this company isn't going to make it, having such a short runway makes even the most brilliant pivot nearly impossible. At that point, if you have the appetite to build something, you might as well start over and build a new business from scratch without all of the baggage.

Option 3: Pivot

Dalton Cadwell, managing director at Y Combinator, shared a tongue-in-cheek piece of pseudocode[8] to determine whether you should pivot or keep working on your product:

```
if (

how well things are working divided by # of
months of concerted full-time effort

is less than

excitement to work on another idea times
confidence you can find an idea that works
better

)

        then {

        you_should_pivot()
```

8 https://www.ycombinator.com/library/6p-all-about-pivoting

}

Sometimes, grit alone won't get you through and you're not ready to quit either. You've tried, you've iterated, but the market just isn't biting. That's when you might consider a pivot.

If Cadwell's pseudocode doesn't make a lot of sense, don't worry, let's just talk about it.

1. **Moderate success despite extensive iteration.** You've launched, you've tweaked the product endlessly, you've pushed it out to many different marketing and sales channels, and yet, still, customer acquisition has been weak or inconsistent, this is a sign that a larger change is necessary.

   In Cadwell's formula, the comparison to make is how well things are going divided by the number of months of concerted full-time effort. This time frame is his proxy for product and marketing iteration. If you're doing just ok after 3 months, keep going. But if you're not much further along after 12 or 18 months, then you should be questioning whether continuing makes sense.

2. **You're still fired up about building.** If you find that despite the struggle of keeping your business alive, you're still passionate about building a massive company and swinging big, then by all means a pivot may be in the cards.

   The Loom founders were knee deep in credit card debt when they went for their second pivot, from a customer feedback platform to an all-purpose video communication tool. As close friends, they had committed to pushing to the very end and were weeks away from running out of money. But that passion and great timing turned into a Product Hunt #1 submission and landed them their first institutional investor, 1517 Fund.

3. **New market opportunities have emerged that capture your interest.** Sometimes in the process of building one thing, you stumble upon another opportunity that seems even more promising. It's like how Slack started as an internal tool for a gaming company. Cadwell's formula multiplies the excitement for working on an idea with the confidence you have of finding something better. You gotta have a hunch that a different exciting new opportunity is out there and we'll talk in Chapter 4 about what that exactly means.

4. **Your team's strengths align elsewhere.** Sometimes with all the product and market iteration you do, you realize that the team you have and the team required to really succeed in the business you're building, doesn't make sense. When the founders of fintech leader Brex were in Y Combinator, they actually were trying to build a hardware VR startup. The founders had no experience in the field, but had built and sold a payments company in their homeland of Brazil. Their pivot reveals an obvious but still valuable lesson: you want to be in a game where you have the home-field advantage. We will talk more about leveraging your assets in Chapter 4.

5. **You have 6-18 months of runway.** Pivots take time, and if your burn rate is too high or your cash reserves are too low, it may not make sense to try to pivot. Of course as Loom, Retool, and others have shown, you can still pull it off. But in an ideal world, you have enough runway to make it work. Which is why starting earlier than you think is important. We'll talk about ways to extend your runway in Chapter 8.

Pivoting is more common than you realize

So, now that you've considered those three options and determined that pivoting is the right choice, or at least decided to learn more about why it might be—let's get into why I'm so bullish about pivots.

First off, pivoting is a pretty common phenomenon and in fact at least one major VC has found it to be more common in the winners of his portfolio than the losers.

Fred Wilson of Union Square Ventures found that at one point about half of their companies had pivoted. In his blog post, he compared "transformers"—companies who made meaningful or complete transformations of their business—to the "stick to the plan" investments that did not make such shifts.

Most notably, Wilson found that nearly two-thirds companies that garnered any sort of return (1x or more) had pivoted while, failures or unrealized (still in-progress companies) were half as likely to do so:

Here's an interesting breakdown of the "transformers" versus the "stick to our plan" investments in my personal track record.

- *5x+ returns – 11 total investments – 7 transformed, 4 did not (63%)*

- *1x-5x returns – 10 total investments – 6 transformed, 4 did not (60%)*

- *Failed – 5 total investments, 1 transformed, 4 did not (20%)*

- *Unrealized – 6 total, 3 transformed, 3 have not (50%)*

Going further back in tech history, one business professor published an analysis in 2000 that analyzed a sample of 400 startups and found that among the successful companies, 93% had to pivot away from their original business strategy[9].

---

[9] Bhide AV (2000) The Origin and Evolution of New Businesses (Oxford

I think part of why you don't see more examples of pivots is that the original business just wasn't successful enough to warrant any attention. You'd never know what they had started out doing unless that story comes out somehow. There's a lot of incentive to make it seem like they've been pursuing this idea from day 1. And the more successful the business is, the less history there is to speak of around the previous business direction.

From recruiting to freight forwarding

As an example of an entrepreneur who is open about his pivot, Deepak Chuggani is founder and CEO of Nuvocargo, a leading logistics company that facilitates trade between Mexico and the United States. His first startup, which he founded after just a few years of work experience at Merril Lynch, was a recruiting platform for (surprise) Wall Street hopefuls called The Lobby. Even after giving it his all for 18 months, the company's numbers didn't look good. In fall of 2018, Chuggani gave himself an ultimatum—if he didn't hit $60,000 in monthly revenue by the end of that year, he would cut his losses and close up shop.

When the deadline passed uneventfully, Chuggani wound down operations and almost immediately started looking for his next idea. Luckily, he was incredibly passionate about building another company. We were actually running tangentially similar HR businesses for a while and pivoted around the same time (December 2018) and I remember his post on the YC forums asking for advice on what to do next.

When I spoke to him for an issue in Mercury Bank's magazine Meridian, Chhugani explained why he was so willing to pivot—he was too invested in this opportunity to back down:

University Press, New York) as described in
https://www.hbs.edu/ris/Publication%20Files/McDonald_Rory_J06_Pivoting%20I
snt%20Enough_0714a59e-a566-4c0d-9005-f425d45913f0.pdf)

"In my case, I was 25. It's scary to think of the energy I had. I didn't come from Big Tech or Harvard or Stanford, and I had gotten into Y Combinator without a technical cofounder. It was like a dream. If I had more of a martyr attitude it would not have worked—but I genuinely saw every piece of it as a cool experience. It was like 15 months into The Lobby and I'm like, fuck it, didn't work but I got into YC, I learned how to build product and raise money. I'm pivoting."

Because he was a single founder, he had more flexibility in what kind of business he wanted to pursue. He knew from his experience fundraising that the size of the market was super important. Raising for a recruiting company was hard, and now he had his eye on something with massive size.

Eventually he landed on cross-border trade—an area where he had some experience because his father had run a small import-export business moving goods out of Asia. It turned out, international mobility was in his blood. Born to Indian parents in Kenya, Chhugani himself grew up in Ecuador and spoke fluent Spanish, making him unexpectedly the perfect founder to build a US-Mexico shipping imperio.

It took months of research and investigation, but eventually he secured a foothold in the market by acquiring a small player in the space and wooed a skeptical cofounder to join after he had really ironed out the details. $75M in funding later, Nuvocargo is rewriting the rules of shipping between the US and Mexico, and Chhugani says that nothing the company has weathered has been quite as challenging as that hard pivot.

To recap: Chhugani's decision to pivot was informed by multiple factors:

1. **Moderate success despite much iteration**—failing to reach the $60k mark he had set for himself

2. **A passion to continue building**—given his youth and his desire to make the most of his YC experience

3. **New market opportunities**—he didn't know right away, but discovering that shipping was ripe for disruption as an outdated industry was a huge reason for his success

4. **Strengths align elsewhere**—his family business and Spanish-speaking chops gave him advantage over other would-be disrupters in the space.

5. **Had 6+ months of runway**—having retained a decent chunk of his $1.2M round, Chhugani actually offered to return money back to his investors. Most appreciated the offer but refused, ensuring he had enough resources to get the new business off the ground.

Don't let your brain psych you out

Stress muddies your mind, blurring your vision of the future. Your brain is wired to keep you alive, not get you to hypergrowth. Cognitive biases ambush your decision-making, pumping up the risks and playing down the rewards of a pivot. But the real risk is sitting on your hands while your business circles the drain.

Loss aversion & local maximum:

No one wants to go downhill. But that's often the only way to climb.

You've found a sweet spot that's working "well enough," and that's your local maximum. It's comfy, and you're scared to lose it. That's where loss aversion kicks in. As YC's Cadwell explains:

"On average, most people take too long to pivot than the reverse. And so, why do they take too long? Loss aversion. When you feel like you've invested in something, you have a really hard time letting it go... You just want to really watch out for a little bit of traction because I've seen that be a trap that capture a lot of really talented people for long periods of time."

As Cadwell points out, founders are hesitant to make changes because you don't want to lose the decent thing you've got going. But here's the kicker: to reach that global maximum—the peak of what you could achieve—you might have to dip down first. It's like taking one step back to take two steps forward. You'll need to leave that comfy spot and risk a little loss to potentially gain a lot more.

Sunk costs & the endowment effect:

It's easy to let old investments and ownership keep you chained to the past, instead moving toward the future.

You've poured time, money, and heart into your startup. That's where the endowment effect kicks in—you're emotionally attached to what you've built. Then there are the sunk costs, the resources you've already invested. You think, "I've come this far, I can't just change course now." But here's the real talk: those past investments shouldn't dictate your future moves. Being too attached to what you've already put in can blind you to new opportunities. Sometimes you've got to cut your losses and pivot, even if it feels like you're abandoning your baby.

Opportunity cost & the multi-arm bandit:

Are you sure this is the best possible option? How can you know?

In the startup world, sticking to one strategy is like a soccer player waiting for the perfect opportunity to take a single shot on goal. You're limiting your chances of scoring big. In reality, professional soccer teams take between 13-15 shots on goal[10] in a match just to score once or maybe twice on average.

YC's Dalton Caldwell, puts it this way[11]:

> *"If you made something and launched it and you think, 'Man, not really working,' a dang good reason to pivot is you get another roll of the dice. You get another shot. And so I've just seen people that use these opportunities really well. It's much easier to be lucky when you get, like, half a dozen shots on goal than one."*

It's like the multi-arm bandit problem from computer science—you have multiple levers to pull (or shots to take), and each one could be the game-winner. But you won't know until you try. So, whether it's tweaking your product, targeting a different customer segment, or even making a more dramatic shift, each pivot is another roll of the dice, another opportunity to collect information and seek a path to that global maximum.

All this to say that we are hardwired to avoid change. You'll probably be able to come up with a bunch of reasons to stay on the same track. But those reasons stem from human biases, the equivalent of psychological pot-holes. They're so ingrained in us that we can't avoid them completely. But we can make smarter decisions by acknowledging that they exist.

Make Your Startup's Second Act Better Than its First

As you've seen, pivoting is more common than you think. You're not alone if your brain's screaming, "Don't risk what you've got!" Here's the truth: sometimes you've got to dip a little to soar higher. Don't let

---

[10]https://bitterandblue.sbnation.com/2013/1/17/3880454/a-look-at-shots-on-target -epl
[11] https://www.ycombinator.com/library/6p-all-about-pivoting

fear or past investments chain you down. You can do better, and it's worth going for it.

Sometimes even dreaming big can be hard; I know because I struggled with it myself. Growing up, I would play the real-time strategy game Starcraft with my high school friends. We would play on "money maps" where the amount of available minerals and gas resources is effectively infinite, a kind of metaphor for venture funding. I would often build a few structures and combat units, but I was easily satisfied with my little squad of troops and failed to invest in more mining units and ramping up my resources. By the end of the game, my friends would steamroll me with huge armies because they committed to playing big.

This was a pattern in my early startups too. I often played it safe, not fully capturing the vision or size of the opportunity. My pitch, products, and results weren't ambitious enough. So unless an investor could see the bigger opportunity in my sometimes mid-sized pitch, I got rejected faster than you can hit "skip" on a YouTube ad.

Don't follow my mistakes friend. Doing a startup means going all-in—no half measures. Your job is to build a war chest, construct a remarkable product that sincerely improves lives, and scale it up as fast and far as you can.

If your current path is filling you with doubt, trust your instincts. Give yourself a real chance to swing for the fences. You might strike out, but at least you played the game like it was meant to be played.

So if you want to pivot like a pro, this book has your back. We'll break it down so you can make moves with maximum confidence. We're going to break it down step-by-step so you can make the leap without the gut-wrenching stress. Stick with me, and your startup's next act will be one to remember.

It's go time.

Recap of Chapter 1: Grit, Pivot, or Quit

You have three options: Grit (stick it out), Quit (yep, literally that), or Pivot (change direction)

- Grit when you haven't exhausted your market potential

- Airbnb pushed through a reputational crisis in the early days

- There is no shame in choosing Quit. Whether in poker (Duke's book Quit) or startups, folding lets you play a different hand

And finally when you should Pivot

- YC's Caldwell's pseudocode for when to pivot

- The bull case on pivoting (I'm biased, but Union Square Ventures data backs me up)

- Nuvocargo case study - how Chhugani pivoted from recruiting to freight forwarding

Fight your brain

- You are programmed to keep working on your current thing because you don't want to give up what you've built. Basic sunk cost math

- See reality for what it is - look through the emotional fog and stress clouds

- You can make it. Don't be scared to dream really big

- Delaying a pivot can waste time to a better opportunity (Cadwell's "multiple shots" advice)

CHAPTER 2

SHOOT THE ZOMBIE

Why founders must find the courage to dismantle their zombie startup and pursue the biggest opportunities

Before we get into your startup's next act, we need to address the current state of your business. This is important so that you start your new chapter with a clean slate, free from the ghosts of the past. The truth is that some of you reading this right now are running zombie startups and are not yet ready to admit it.

Elle Morrill and her cofounders launched Referly out of Y Combinator[12] the same summer we posed for that fateful Vanity Fair photoshoot. The company helped bloggers make money from affiliate links to third-party products and a few months later announced they had 10,000 users and were growing 45% month over month.

---

[12] Fun fact, I wrote a recommendation for her YC application

But less than a year later, Morrill announced that her company was pivoting in a blog post[13] that coined the iconic phrase "zombie startup".

More than failure, she said that she feared Referly would turn into Bruce Willis in *The Sixth Sense*—a company that is ostensibly alive and operating "but might as well not be".

In the face of that fear, Morrill shut down the affiliate commissions part of their business to double down on their pay-per-post model, hoping to defibrillate their flatlining business and give her team a chance at a big win. As she wrote in her post, it can take a startup a long time to die:

> *"It has been 6 months since [Y Combinator's] Demo Day and I don't think anyone has officially died. So I'll say it. Referly died. It's not the kind of dead where the website goes dark and everyone gets jobs somewhere else. But the idea that we started with turned out to be the wrong one, so we killed it and yesterday I acknowledged publicly to ourselves and everyone else that we have to change our course."*

Morrill decided to make the pivot because she wanted to pursue a bigger opportunity and she was willing to risk what she had built so far to get it.

After a more iterative pivot fell short, Morrill and her cofounders tried again, reinventing their company into a data platform for private companies. The newly fashioned Mattermark became a go-to resource for investors to discover and vet potential deals. Their blog was read by many founders, investors, and senior operators across Silicon

---

[13] https://www.daniellemorrill.com/2013/03/zombie-startups/

Valley. After raising a Series A and B, the company was acquired by Full Contact in 2017[14].

The Dangers of Becoming a Zombie

Startups begin with the spark of wild ambition and deep conviction.

You quit your job and take the leap into the fire of the unknown. Building furiously into the night. Hustling to close early customers, launch V1 of the product, patch a hotfix to production, all the while dreaming of becoming the next Collison brother (Stripe), Melanie Perkins (Canva), or Tope Awotona (Calendly).

But after a while, maybe six months, maybe two years, things start to change. The company gains some recognition and respect. Investors and team members have expectations for what the company is going to do next. And suddenly you become a lot more risk averse. There's more to lose.

Decisions are taken slowly, and putting them into action takes even longer. Growth grinds to a halt or creeps forward at a glacial pace. Admitting you might be a zombie company is hard. But the alternative, staying in denial, is worse.

My cofounders and I spent six months trying to execute a pivot for Ridejoy. But after countless brainstorms and discussions, we were unable to converge on a new direction.

At the time we might have said we were trying to find the perfect idea. That it was better to do nothing than to pursue something that wouldn't work out.

The truth is, we were scared. Afraid of failing, of disappointing our investors, of disappointing ourselves.

---

[14] Sadly the Mattermark acquisition fell short of many people's expectations, but I still consider their pivot a success. It's just another reminder of how hard and rare it is to make it as a VC-backed company

As long as we didn't make any big new decisions, we could tell ourselves we were still alive, still going, still in the fight. But we were actually dead and hadn't admitted it to ourselves.

It's easy to forget how rare it is to build a breakout startup. According to a 2019 analysis of 2,622 seed investments made on AngelList, a startup that has raised a seed round of venture funding has an estimated 1 in 40 shot—or 2.5% chance—of achieving unicorn status[15].

> *"We're in the fluke business. The whole point is to get the fluke."* — *Marc Andreessen, a16z*

Taking venture capital means you're aiming to be the 1 in 40. It's the combination of rapid growth and massive scale that make the economics of venture capital viable.[16]

Now you've lowered your ambition because you've realized just how hard this game really is, and you're trying to maintain some level of conviction that you can succeed, even if at a smaller scale. But that's not the right way to think about it.

The Courage to Stay in the Fight

If you think you're running a zombie company, then you need to reset your ambition and rediscover your conviction. Brace yourself. This bitter medicine won't taste good, but it might just be what you need.

Here's how you face the prospect of a stalled out startup:

Confront reality. Morrill suggests the following questions in her article that might help you come to terms with the fact that your company is dead on its feet:

1. You don't want to get out of bed in the morning

---

[15] https://www.angellist.com/blog/angellist-unicorn-rate
[16] See appendix for more details

2. You don't want to go out in public for fear you'll have to explain what you do

3. You haven't hit 10% week-over-week growth on any meaningful metric (revenue, active users, etc) at any point in your company's history

4. You're working on the same idea after 12+ months and still haven't launched

5. You've launched a consumer service and have less than 2% week-over-week growth in signups

6. You've launched an enterprise service and have less than 2% week-over-week growth in revenue pipeline

7. You are the CEO and hole yourself up in the offices so you don't have to talk to employees and can read TechCrunch

8. You've hired consultants to figure out revenue, culture, or product in a company of less than 10 people

More than a decade after she published it, Morrill's list is still a great starting point for evaluating your business's zombie status. But in truth, there is no specific criteria that can tell you whether you have an undead company on your hands—you have to feel the truth in your gut.

Remind yourself of why you started this. No one does a startup because it was the prudent thing to do. Sure being your own boss might motivate any entrepreneur (startup or otherwise) but I'm guessing you were compelled by one or more of the following motivations:

1. You've invented a new technology, product, or service and want to bring it to the world

2. You wanted to rapidly develop your skills and knowledge by working outside of a corporate structure

3. You're trying to make a massive impact on people, companies, and society through your company

4. You want a shot at a life changing sum of money

Take a moment to think about it and you'll realize that all four of these possibilities are only doable if you actually take a dramatically different tact and pursue a bigger opportunity. You won't have the runway, brand, or capacity to hire and work with great people, develop something amazing, reach millions of customers, and potentially cash out big if you keep on your current trajectory. It's only possible when you take a leap.

Rekindle your conviction in yourself. You've been beaten down by the market, your competition, and your technical woes. But as bad as you feel, you're actually more capable than ever. Think of all that you now know and can do that you couldn't when you started. So your business isn't where you want it to be—you are not alone.

The industry is full of founders who were on the brink of failure, who seemed wildly unimpressive, and who found an unlikely way to the top.

Today Airbnb is one of the crown jewels of the YC portfolio but in 2007 the founders had to sell custom cereal boxes styled after US presidential candidates of the time (Obama O's and Cap'n McCain's) scraping together $30,000 to fund their startup ahead of any seed capital. More on their journey in the next chapter.

But then there was the birth of Pixar.

Originally a specialized team within Lucasfilm, they found themselves at risk of being disbanded due to George Lucas's post-divorce money troubles. Ed Catmull and Alvy Ray Smith took it upon themselves to secure funding, but were rejected by 35 different VC's. They even came close to striking a joint deal with General Motors and Philips, only for it to fall apart at the eleventh hour.

At this critical juncture, they reached out to Steve Jobs, who had recently been ousted from Apple. Seeing the potential, Jobs invested $10 million to launch Pixar as a standalone company. This pivotal moment not only saved the team but also set the stage for Pixar to revolutionize the world of animated films.

More on Pixar's pivot later in this book, but if the world's greatest storytelling company and the source of Jobs's fortune[17] was almost dead on arrival, who's to say you can't still pull this off?

Understand that failure is not final. When we were struggling with Ridejoy, our friend Andrew Lee, then a recently exited founder, now investor at a16z, took us out to dinner (at the Zynga offices, hah). He told us that we should realize that the most likely outcome of our startup was failure. "So just ask yourself 'What kind of crazy shit do I want to try before I go under?',", was his advice.

Remember that before it was called Venture Capital, it was called "Adventure Capital'" because investors were putting money into things that no one else would dream of[18]. Failure is not just common, it's mundane. Most founders get second, even third chances, especially if they stay bold and aggressive in their earlier attempts.

Life is Short, Don't Waste It

Life is shorter than we think and spending that time on things that don't matter is a poor use of your finitude and your investors' money.

---

[17] When Disney bought Pixar for $7.4 billion in 2006, Jobs became Disney's largest individual shareholder. That deal was worth way more than his Apple shares, making Pixar the real moneymaker for him.

[18] One forward thinking investor of such risky businesses sought to distinguish himself from those boring New York investment banking firms. John Hay Whitney and his colleagues chose the term "private venture capital investment firm", shortening the term "adventure capital" which was already in colloquial use, when speaking with the New York Times in 1947. *The Power Law: Venture Capital and the Making of the New Future*

As Morrill wrote in her 2013 blog post:

> *The biggest reason to charge ahead is that I don't want to waste a single moment of my life in denial, in deadlock, in zombie mode waiting for something I can't control or change or expecting magic to happen. ... I simply can't, won't, would never give up precious days, weeks, months, years. And it's not that I don't have endurance for the schlep, but I can only summon that super-human power to fight for the right thing.*

Morrill's Mattermark was successful for a number of years, but eventually folded in a bit of a disappointment. But none of these things really mattered—she went on to become a successful angel investor, her cofounder Andy Sparks went on to found a highly-respected publishing platform, Holloway, and life went on.

When we returned nearly half our seed round back to Ridejoy's investors, some appreciated this, at least when speaking to our face.

"Better to get 50 cents on the dollar than nothing," they told us.

But others were more judgemental. What the hell were we doing, wasting their capital and their time by giving up halfway? Most of their investments don't succeed and even the ones that do have very little effect on their overall return.

This is a difficult fact to confront, even for longtime investors who have seen the principle play out again and again. Paul Graham of Y Combinator writes[19]:

---

[19] https://www.paulgraham.com/swan.html

"It's hard to make ourselves take enough risks. When you interview a startup and think 'they seem likely to succeed,' it's hard not to fund them. And yet, financially at least, there is only one kind of success: they're either going to be one of the really big winners or not, and if not it doesn't matter whether you fund them, because even if they succeed the effect on your returns will be insignificant."

So go ahead, shoot your zombie. Accept that failure is possible, even likely. But remember that you might beat the odds, and your best shot at doing so is by swinging big.

Recap of Chapter 2: Shoot the Zombie

Facing your zombie (startup)

- Morrill's Referly → Mattermark (Bruce Willis in *6th Sense*)
- Check the signs that you're already a zombie: slow/no growth, still haven't launched, spend all day doomscrolling
- Andresseen: VC-backed startups = the fluke business

Finding hope and new life

- The story of Pixar's pivot (Lucas's divorce → high end computers → 3D movies)
- Ridejoy failed because our team was afraid of failing again
- Life is short, don't waste it. PG: "It's hard to make ourselves take enough risks"

CHAPTER 3

PILOT YOUR PIVOT

Use the Align-Explore-Commit framework, to test-drive your pivot before you go all in—reducing uncertainty and increasing conviction.

After the floundering Ridejoy pivot, I learned a hard lesson about the power of structure and deadlines.

When we were pivoting my tech hiring startup Headlight into Midgame, a set of AI-powered tools for gamers, we started with an unexpected deadline: our employee left for a conference, creating a compressed week of research and decision-making that pushed us to decide to pivot by the time she was back the next Monday. But that was just the beginning.

We had very little runway and needed to raise to continue building out our vision, but few investors were willing to take a bet on us or our pivot. So despite having already raised a pre-seed round and being "past" the accelerator stage, we applied to and got accepted to Alexa

Accelerator, a joint venture between Techstars and the Alexa Fund, an investment vehicle associated with Amazon's voice-enabled assistant.

While in some ways, it felt like a step back, the program was critical to our success. We had a 10 week deadline. My cofounder and I moved into tiny college dorm-style rooms in Seattle. Every week, we presented our progress and metric movements at an end-of-the-week performance review held with all of the startups in the batch, and just a few months later we strode on stage at Demo Day with a powerful narrative in front of hundreds of investors.

You need structured accountability to navigate the uncertainty of a pivot. This chapter shows you how to create that for yourself, with a Pivot Pilot—that structured period of exploration before fully committing.

Pivoting is a very difficult thing because it involves admitting that your prior conviction was mistaken for some reason or another. While that's totally normal and expected in a startup, it still needs to be handled with care.

If you just sort of wing it, some of the people who believed in you—your cofounders, investors, team members, customers—may feel they can't count on you any more and stop trusting you, which could quickly erode whatever assets you still have (brand, team, funding, etc).

It can also be hard because you lose a lot of the momentum that kept the business humming along. The cadence of team meetings, new customers, upcoming launches, etc is lost and it can be easy to lose the sense of urgency that you desperately need to make the transition successful.

What you need is a structured way to move from losing conviction in one idea to gaining conviction in another.

What you need is a Pivot Pilot.

Maintenance Mode

A key part of getting the Pivot Pilot right is dealing with the startup's current business. You can't go deep on pivot exploration if you're spending your days fire fighting. Take your foot off the gas pedal and switch on cruise control. Putting your business on maintenance mode means pulling the plug on all high-intensity efforts to grow. Actionably, this means turning down the marketing noise and hitting pause on the development of new product features. Having your business running on autopilot gives you the space to explore new horizons.

Let's take the example of Lyft. The company started out as a carpooling platform branded as Zimride—we actually competed with them for a while in the long-distance ridesharing market with Ridejoy. The founders decided to experiment by launching an intra-city ridesharing mobile app while still running the carpooling play. As they saw traction with their new idea, they slowly phased people onto the Lyft work and off Zimride, eventually selling the revenue-generating business line to Hertz only after they had gained significant traction and total conviction in the ride-sharing space.

The idea of maintenance mode also means that you might have an opportunity to return to the original business. This reduces the sense of loss aversion and sunk costs, because you don't give it up right away. And you may legitimately find that the business thrives without your direct oversight and becomes something you return to with new insight after the Pivot Pilot.

But more likely the business will suffer—which is to be expected and not a bad thing because it makes shutting it down in the end less painful.

The importance of a hard reevaluation deadline

The early days of a pivot are like a breath of fresh air. Like someone threw open the window of a musty old beach house. You're feeling like a kid on the first day of their summer holiday with a pack of crayons and a blank piece of paper. The second chance comes with huge potential for upside, and you're itching to get going. The whole team is sprinting forward on the adrenaline of starting over. New beginnings. It's exciting, heady stuff.

There's one problem though. You don't know which direction to channel the energy in.

Your Pivot Pilot is a time for exploration i.e. finding a problem people will pay you to solve, and then using available resources to build the solution. Getting through this phase can be a little tricky because there's no clear path forward. It's easy to veer off course and get lost racing down a bunch of rabbit-holes. It's also easy to get overwhelmed by the endless options and be paralyzed into doing nothing.

Timelines are what give the pivot exploration phase structure.

A definite end-date will push your team forward at a time when they won't see the usual markers of progress like an uptick in MAU or new sales orders. It is a powerful tool to drive focus and momentum.

Choosing a reasonable timeline is key. Take a temperature check of your human and financial resources. Ask yourself the important questions — how fast can your team adapt to a new idea? Are they motivated to keep going? How much runway do you have left?

Stay firm to the deadline

One thing I've seen a lot of founders do is set a squishy deadline and then keep pushing it out when they aren't ready to make a decision.

Don't do that.

It's important to make the deadline feel as real as possible, and that can mean leveraging external accountability.

You can give weekly updates on the pivot on LinkedIn like Multi, who we will talk about later in the book, or even just invite friends and advisors to a designated Zoom call. Make sure you're accountable to the deadline.

With Midgame, we had both Techstars and Betaworks Camp Demo Days that spurred us on. And of course the pressure to deliver on YC Demo Day is legendary. Some founders even try to keep the pressure on by having a post-YC sprint to keep their productivity up.

Before diving headfirst into the exploration phase, put an evaluation mechanism in place. If all goes well, you're going to have a bunch of ideas on your hands at the end of this process, and having a set of objective measures to test them will help you make an informed decision. Make sure you have a healthy mix of quantitative and qualitative metrics.

The Align-Explore-Commit framework

A Pivot Pilot is a 4-8 week period after you've realized you might be operating a zombie startup, and before you've gone all in on a new direction. Remember, no Hail Mary's!

By using a Pivot Pilot, you increase your conviction with your new direction, reduce thrash and possible blowback from your stakeholders, and give your new company the best chance of success. A Pivot Pilot follows a three step process of Align-Explore-Commit:

1. **Align**—you start by getting everyone on board with the idea that your company direction isn't working and a more significant change is warranted via a structured period of exploration. We'll talk more about having those conversations in Chapter 7: Align Incrementally and how to

make sure you have the cash to pull it off in Chapter 8: Extend Your Runway.

2. **Explore**—you put your core business on maintenance mode and set a designated time frame to research new markets, build and test prototypes, and gain conviction in a new direction. We'll talk more about other aspects of exploration in Chapter 4: Leverage Your Assets, Chapter 9: Get Out of the Building, Chapter 10: Narrow Down, and Chapter 11: Work Like a (Mad) Scientist.

3. **Commit**—you gather your stakeholders to make a final decision about whether to pivot or return to the original business, allowing everyone a chance to feel bought in. We'll talk more about the decision in Chapter 5: Make the Call, how to sell it in Chapter 6: Own Your Story, and how to keep going when things get hard in Chapter 12: Stay Resilient.

This approach works because you bring people along and get their buy in on the final decision by distributing shared awareness of the problem. It also works because it preserves your optionality temporarily through leaving your existing business on maintenance mode.

A sample timeline for a pivot

Here's a high-level summary of how this process works with a rough 10 week timeline (yours could be shorter or longer depending on the size of your team and what you discover in your exploration process).

ALIGN

Week 1: Instead of announcing abruptly that the company is pivoting, you start by noting your concerns with your current premise, based on what you've learned since starting the company, and proposing a period of exploration to explore new opportunities and shore up weaknesses in the current premise.

Week 2: You align incrementally with all your key stakeholders — founders & leadership team, investors/board, full team, and possibly customers to make sure everyone is bought in (or at least informed).

EXPLORE

Weeks 3–7: You place the business on maintenance mode and run a 4 week exploration period where you spend 60% of your time on studying a new area, narrow in on a problem and target audience, and run experiments in potential solutions and marketing channels. We'll cover this in later chapters in detail.

Week 8: You build a presentation or memo that summarizes what you did and what you learned, and what you now conclude you should do. The options are:

A. Pivot into a new, more promising area

B. Continue running the existing business armed with some learning

C. Kick off a follow up exploration period

D. Shift into one of the pivot alternatives (sell, shutdown, etc).

COMMIT

Week 9: You realign with your stakeholders—this time with your presentation and set of recommendations. Work through any disagreements or issues. If there are fundamental shifts to what you are offering your customers, you will need to come up with a plan to phase them out or help them transition to a new provider. But chances are, they weren't really using the old thing that much and will also potentially be excited about your shift.

Week 10: You move forward with whatever you decided with full alignment and buy in from your key stakeholders. If you are pivoting, begin the process of winding down the old business.

Mistakes we made with Ridejoy

As I mentioned in the introduction, our pivot with Ridejoy failed—for a variety of reasons—but many of our self-inflicted problems arose from how we handled the initial decision to pivot.

Flubbed the Alignment step. We literally just announced to the team a few days after the Craigslist cease & desist that we would need to pivot, and tried to enter the exploratory period immediately. We didn't make a communication plan about why this was a problem and how we'd solve it or give our investors a heads up until much later. We'll talk about how to do that well in Chapter 7.

Made the Exploration period too short. We gave the team a mere two weeks to work on new ideas, without much structure or process. Granted, we asked people to work on it full time instead of half time, but in some ways that made it even more jarring. Then we abruptly told everyone "We don't like any of the ideas we came up with, and we can't afford to keep everyone. So, we're laying you all off in order to preserve cash to come up with a new idea." We'll talk in later chapters about how to structure your exploration period effectively and communicate better with stakeholders about your decision-making process.

No deadline to Commit. We didn't timebox our next founder-only exploration period, which meant we spent the next eight months pitching different ideas to each other with no clear deadline or next milestone. This is also when we finally told our investors we were exploring new ideas (without much of a hypothesis on what we wanted to do) and one big investor did try to ask for the money back. We eventually convinced them to stand down, but that was stressful.

You now have the strategy for making this pivot work—set up your check in date and get going. If you have cofounders or investors who are not fully bought in, jump down to Chapter 7: Align Incrementally, to learn how to win them over.

But as we enter the Explore phase, I need you to make sure you are capitalizing on your unfair advantages and riding the biggest trends. Because that's how you get exponential growth. The next chapter will show you how to do just that.

Recap of Chapter 3: Pilot Your Pivot

Use the Align-Explore-Commit Framework

- We don't do Hail Mary's. Use a 4-8 week structured approach to your pivot

- Ridejoy's downfall - endless exploration can kill you

Having a deadline and *sticking* to it is critical

- Set a hard date and build ways to stay accountable to it

- Landing the hard pivot with Headlight - an unexpected conference, weekly metrics review at Techstars

The Maintenance Mode strategy eases the transition

- Lyft case study - kept enterprise biz (Zimride) running while incubating new model (eventually Lyft)

- Scale down, don't shut down (right away at least)

- Be decisive but also considerate - pivoting is hard on you and your team

CHAPTER 4

LEVERAGE YOUR ASSETS

Identify and build on top of your existing product, audience, insights, and position in the market to move on trends and create new opportunities

A pivot only makes sense because you are leveraging something from your existing business. So let's take the example of the basketball pivot.

Picture LeBron, Curry, or your favorite NBA player in action. He's got one foot firmly planted on the ground, swiveling, scanning the court as he decides his next move. It's kind of the same with your startup pivot. Sure, you're going to give up a bunch of the progress you made on your original idea. But hey, you've still got one foot on the ground. Start thinking about the things you can hold on to. More importantly, what you can actually use.

Your company will have to have built something by this point. Maybe it's a tech platform, maybe it's an audience, maybe it's an understanding of the problem or an operational system that is highly effective. Ideally, one of these elements can be a focal point for your pivot.

You need to have an "unfair advantage" against anyone else who is in this space. I used to find this idea difficult, because I believe in fairness and it seemed like saying "you need to have a way to cheat the system" or "you need unearned privilege". But over time, I've come to understand this really refers to choosing a path where you will naturally have some ability to do better than competitors. A naturally tall person is more likely to succeed as a basketball or volleyball player than as a gymnast or jockey. It's the same idea. What has your company done well that seems comparatively more difficult for others to do?

I recently saw a tweet of a young founder asking what the best way to build a private community of highly successful entrepreneurs would be if they hadn't already done that. The answer in the replies was "you don't, you first have to build something else successful." Some businesses can only be started by someone with a certain background[20].

Assets & insights

The things your startup has accumulated so far can be broadly split into two categories: assets and insights. Both these concepts are best illustrated by examples from companies that have pivoted.

Asset: Loom's first product was a user-testing tool for startups to get smart feedback on their prototypes from experts. The user-testing product never picked up and the founders had as little as 2 weeks of runway left when they started thinking about pivoting. The team

---

[20] Sam Parr could not have built Hampton (a vetted community of high growth founders) before he started The Hustle and sold it to Hubspot. https://x.com/thesamparr/status/1640680965830762496

realized that some people were using the tool for something Loom never intended, communication.

Capitalizing on this, they carved out part of the product and launched a dedicated screen and video recorder. The asset Loom used for their pivot was part of their original tool. They leveraged existing technology to satisfy new customers (work colleagues) and solve a new problem (communication). They got acquired by Atlassian for just under a billion dollars, so it's safe to say that the pivot worked out!

Insight: When Discord founder Jason Citron saw the rising popularity of iPads, he thought he had spotted a golden opportunity. Citron wanted to build a startup that made multi-player video games designed for tablets (as opposed to PCs).

Their first game, Fates Forever, got great reviews, but tablet gaming just wasn't catching on. In the process though, the struggling startup noticed something interesting — users seemed to really like the inbuilt communication features in Fates Forever. When they dug deeper, they found that most gamers at this point were using Skype or TeamSpeak to communicate and they all hated it. Citron and team doubled down on this valuable insight and ended up with Discord.

What wave are you riding?

I know we've been talking about stuff that your company has actually built up to this point, but the reality is that startups don't exist in a vacuum. They are living, breathing entities that operate in a dynamic world that's always in flux. There'll always be a bunch of external factors at play which have a pretty big impact on your success.

PG has a great line about this[21]: "Every company that gets really big is "lucky" in the sense that their growth is due mostly to some external

---

[21] http://www.paulgraham.com/convince.html

wave they're riding, so to make a convincing case for becoming huge, you have to identify some specific trend you'll benefit from. Usually you can find this by asking "why now?" If this is such a great idea, why hasn't someone else already done it? Ideally the answer is that it only recently became a good idea, because something changed, and no one else has noticed yet."

What's really neat about trends is that they level the playing field. Unlike assets and insights that rely on stuff your startup has already done, trends are out there for everyone to see and jump on. But here's the catch — since it's a free-for-all, it's important that you try and get in on the trend early. You've already got a pretty big advantage over the newbies because you have an actual operating company. You've got the infrastructure all set up to get straight into it.

Think of it like a game of soccer. You've just sprinted halfway down the field, and now someone's headed the ball back in the opposite direction. You're all warmed-up, heart pumping, and breathing heavy. You don't slow down. You turn around (i.e. pivot) and keep going. Now imagine someone who just got off the bench. They're a little chilly, still figuring out the game's rhythm, maybe a tad disoriented. Even if both of you spot the ball at the same time, you'd still get there first.

And getting there first matters. You gotta think about why right now is the best moment to build your product or service. Take Jeff Bezos founding Amazon just as the world was on the cusp of the internet era. Bezos talks about it in an interview at the Special Libraries Association Annual Conference in Seattle circa 1997[22]:

---

[22]https://www.businessinsider.nl/1997-jeff-bezos-amazon-empire-viral-video-books-2019-11/

"Three years ago I was in New York City working for a quantitative hedge fund when I came across this startling statistic that web usage was growing at 2,300% a year, so I decided I would try and find a business plan that made sense in the context of that growth..."

To be clear, most people aren't as proactive as Bezos. He also had unshakeable conviction in his premise (the growth of the internet). Like we discussed way back in Chapter 2, developing deep conviction and then acting on it is a fundamental key to startup success.

We all scroll through tons of stats on Twitter/X or LinkedIn or whatever social-media platform which happens to be in favor when you read this, but 99.9% of people don't start a company around it, and only a small handful have come close to Bezos's level of success. And he's not even an engineer!

But this isn't about Bezos. It's about you, your company, and your pivot. The truth is, a pivot is a great time to capitalize on the hottest new trend in the market. Riding a trend is like surfing a wave. If you're lucky enough to catch a green wave (one that hasn't broken yet, but is just about to) at the right moment, you can ride the momentum all the way to a win.

The six kinds of trends

So, how do you find a good trend? The truth is, trends are unfolding around us all the time. You just have to know what they look like. Here's how I'd divide them up:

1. **Technology**: A new technology that's changing the way companies and people live, work, and play. The most obvious example as of Spring 2024 is Generative AI, specifically large language models, image diffusion, and sound synthesis.

   Historically, these trends come in waves. It started with smaller, more powerful microprocessors enabling the personal computing era. Then, as internet connections got cheaper and faster, more and more people used their PCs to go online. Just like the microprocessors, the PCs kept getting smaller and more powerful. Now most people have pocket-friendly mobiles that are way more useful than the bulky PCs of the '90s: what with internet capabilities, GPS, and high-quality cameras. Plus millions of apps that let you play games, talk to friends, get information, do work, and express yourself.

2. **Consumer behavior**: There's a noticeable shift in the way people are living, working, or playing that opens up a new opportunity. This happens when people start operating differently — leading to a tangible change in what they desire and how those desires can be met. Think the shift from getting takeaway to meal kits, or working out in gyms to having a fitness-at-home setup. To survive, businesses need to have dynamic responses to evolving lifestyle choices and preferences of their customers.

3. **Distribution**: A new way of reaching consumers or businesses, often powered by a mix of rapidly adopted technologies along with new consumer behaviors. This is something like companies going from reaching users only through their websites to also building Chrome extensions or using messaging platforms to sell ecommerce via DMs and chatbots. There's a clear shift toward engaging with customers through more direct and personalized avenues.

4. **Business model**: The way people acquire, purchase, or pay for solutions has changed — and your business model is built on the new way that things are handled.

The most prominent example of this is the shift from packaged software bundles to SaaS. Of course, subscription-based models have become popular beyond software — from monthly pet product subs (BarkBox) to exclusive wine-a-month clubs. Another big one is APIs. This model allowed highly technical teams to sell usage-based services to technical customers by simply implementing a few lines of code—a developer's dream come true, and an approach that was unheard of only a few decades ago.

5. **Operational**: The way businesses generate or deliver services has changed fundamentally, thanks in part to new technology. A great example of this is DoorDash— back in the day, restaurants would have a FT employee dedicated to completing food deliveries. But that has been completely replaced by gig-economy workers that contract with DoorDash, Instacart, Uber and others. Businesses like this are leveraging the power of technology and flexible labor markets to optimize operations.

6. **Regulatory**: A new set of laws, regulations, or deregulations can often create new problems or opportunities for entrepreneurs to capitalize on. For instance, crowdfunding platforms like Republic are built on the 2012 JOBS Act that allowed the general public to invest in startups under certain conditions. Meanwhile the patents for Rogaine and Viagra expired in 2019 and 2020, allowing new startups like Hims, Romans, and Bluechew to sell these health products to men discreetly over the internet, both a regulatory and distribution (telehealth) trend.

How GOAT and Retool Did It

GOAT—the sneakerhead haven

The company started out as Grubwithus, a startup based around helping adults who have shifted to new cities make friends by organizing group dinners. 4 years into building Grubwithus, co-founders Eddy Lu and Daishin Sugano realized the business wasn't scalable. They stumbled onto the idea for GOAT when Sugano, a bonafide sneakerhead, had a bad experience ordering Air Jordans from Ebay. He got fake kicks and couldn't get his money back. Sugano wasn't happy, but something else caught the co-founders' attention: the growing number of serious sneakerheads. Premium sneakers weren't a geeky, niche interest; they were becoming a mainstream status symbol.

Lu and Sugano saw a lot of problems in the young market[23]. Finding a pair of sneakers in the style and size you wanted was incredibly difficult. You had to get onto ebay and scroll through hundreds of listings posted by different sellers — comparing prices, styles, sizes, shipping costs. The worst part? Even after doing all that work, there was a high chance the shoe you paid a lot of money for was a fake. The only other option was going to one of the very few brick and mortar stores, but the shoes there were expensive and they typically didn't have a lot of inventory on-hand.

When Lu and Sugano built GOAT, they changed the shopping experience completely. They created a solid database of sneakers and then built a searchable site to help customers find the SKU and size they needed at the lowest price. Key trends they jumped on:

1. **Consumer behavior**: Noticing that more and more young people were passionate about collecting premium sneakers.

2. **Distribution**: Building a dedicated sneaker marketplace and putting a screening process in place before a seller could post a listing.

3. **Technology**: Using image recognition tools that could greenlight if a pair of sneakers were authentic from 7 specific pictures the seller was instructed to take.

---

[23] https://medium.com/for-entrepreneurs/ready-set-goat-2bf109902e37#.rkpkp0lrq

With strong execution, GOAT joined the coveted ranks of the unicorn club, closing their last funding round at a valuation of $3.7 billion in 2021.

Retool—helping builders build

Retool started out as a YC-backed fintech company hoping to make it as a Venmo competitor based in the UK. But things didn't go as planned. A few weeks before demo day, the startup was bleeding $1,000 per day. Founder David Hsu had less than 60 days of runway left when he was forced to think up a new direction for the company.

Luckily, Hsu didn't have to look too far. His fintech startup had a bunch of internal tools to comply with sticky fintech regulations. These tools had been an absolute pain to build, but the basic stack for all of them looked pretty much the same. Hsu saw more and more devs even in other companies "wasting" precious time building out internal tools — so he decided to build a SaaS platform that would help them build internal software faster.

The trend Retool jumped on? Increasingly complex internal tooling. More and more software companies were building and using internal software to run their businesses more efficiently. According to a research report drafted by Retool[24], more than half of all applications in the world are internal tools.

As of 2023, Retool is cash flow positive. Their customers include prominent Fortune-500 companies and they've raised nearly $200 million from investors like Sequoia and PG at Quiet Capital.

Backing the wrong horse

A word of caution: Obviously, you can win big if you leverage the right trends, but trends are a very hit or miss thing. You talk to veteran technologists and you'll know that for every desktop→mobile and

---

[24] https://research.contrary.com/reports/retool

Generative AI shift, there are dozens of technologies that have not lived up to the hype, yet[25].

- Web3 technologies (DAOs, NFTs)

- Internet of Things (IoT)

- Personalized medicine

- Quantum computing

- Autonomous vehicles

- 5G internet

- 3D printing

Trust personal experience over projections or "big data".

When we pivoted from Headlight to Midgame, we started out pitching the rapid growth of esports as a hidden trend that we were riding with an analytics tool for gaming teams.

The only problem?

While both my cofounder and I were avid gamers, neither of us were truly passionate about the competitive esports team, hadn't attended any games, and had no friends or contacts within the exclusive world of professional gaming.

We relied on the data and analyst reports about the rise of esports. But it turned out most savvy investors and truly fanatic entrepreneurs had founded companies in this space 2 or 3 years ago.

We were too late—the hype train had left the station. And in the end, esports did not transform the sports / entertainment industry anyway.

I'll close this chapter with another gem from Bezos:

---

[25] I write "yet" because you truly never know. While the Internet was overhyped in 1999-2001, it has since completely transformed human life and generated trillions in value. I'm sure I'll be wrong about some of these.

"The thing I have noticed is when the anecdotes and the data disagree, the anecdotes are usually right. There's something wrong with the way you are measuring it. And it doesn't mean you just slavishly go follow the anecdotes then. It means you go examine the data because it's usually not that the data is being miscollected, it's usually that you're not measuring the right thing."

Meaning trust your own eyes and ears over 3rd party measurements. If your customers are rapidly adopting a technology or platform or your fellow founders are rapidly acquiring users with a new distribution channel, that's far more honest a signal of a real trend than any report Gartner might put out.

☀

Another place where going with your gut is crucial is actually making the final call around your pivot. Your exploration period is going to go by fast, and as you barrel towards that decision date, you may feel under the gun.

"How can I make a call on this?" you wonder. "I just got started!"

Well, no decision is ever made with full information, and doubly so for a pivot. In the next chapter I'll show how to get to conviction even without all the data.

Recap Chapter 4: Leverage Your Assets

Launch your new pivot off of built assets and earned insights

- Loom case study on how to get the most out of your existing product
- Leverage "insights": the story of Discord's gaming founders

Find a trend to ride on

- Bezos went and risked it all on the internet in 1997—what about you?
- 6 types of trends: tech, consumer behavior, distribution, business model, operations, regulatory
- GOAT case study - riding the sneakerhead wave
- Retool's founders went from fintech over to internal tooling
- Bet on your gut not spreadsheets

CHAPTER 5

MAKE THE CALL

*Use data and discussion to get an informed perspective
about your pivot—then go with your gut*

Nothing clears your mind like thinking you're probably going to get
fired. At least that's how the founders of Intel felt.

Founded 1968, the company's first hit product was actually in
memory chips, or RAM. Their 1103 chip was the best-selling
memory chip in the world in 1972, and Intel dominated the space for
most of the '70s. But by the early '80s, Japanese companies were
eroding Intel's market dominance with capable and competitively
priced chips of their own.

The team at Intel spent years embroiled in company-wide debate
about how to respond to this slow motion disaster.

Some wanted to invest in their memory chip business, others wanted
to target niche markets, while others wanted to focus on their newer
product line, the microprocessor, which was only a fraction of their
total revenue.

In 1985, with memory chip market share down 97% and their U.S. competitors going under left and right, cofounder Andy Grove and CEO Gordon Moore had a crucial conversation in one of the office cubicles. As the story goes, Grove mused that they might get replaced by the board with a new executive. "What would the new guy do?" they wondered.

The answer was clear: They'd get Intel out of the memory business.

So the two decided to make that decision themselves. The decision had painful consequences: Moore and Grove laid off a third of their workforce (7,000 employees) and chose to double down on the nascent CPU business. From January 1985 to January 2000, the stock price grew by 99,300%, and today Intel is still the dominant microprocessor manufacturer in the world.

Pull the plug, take the plunge, cut the cord—call it what you will, this chapter is all about making that decision.

Getting a handle on the options

At the beginning of this book, we talked about your choice to either Grit, Pivot, or Quit. Now that you're in the process of pivoting, you have three options — Commit, Return, or Keep.

1. **Commit** - You go all in on the pivot. You are choosing to shut down, sell, or otherwise completely stop investing in the old business. This doesn't mean you have everything figured out, it means you have more conviction that this direction can yield a chance at a billion dollar outcome versus your old path.

2. **Return to the original business** - This means giving up on making a major change, and choosing to stay invested in the

original business. Perhaps with a new insight or feature that you've developed during this pivot pilot. This makes sense if the original business continues to thrive without you, your efforts to pivot are really not good, or some new trend emerges.

3. **Keep exploring** - This means you stay in the pilot pivot for longer. Fair warning: This is dangerous to do, and only makes sense if you are genuinely unable to decide between Committing and Returning because the old biz is still doing well, but the new area also looks really promising.

When to pull the trigger

The next thing you need to figure out is knowing when to make this call. Pulling the cord is tough because it forces you to balance two valuable things — information and time. It's quite straightforward. The more customer interviews you do or MVPs you ship, the more information you get (good) + the more time you spend (bad).

Brian Christian and Tom Griffiths (the computer science authors behind the book *Algorithms to Live By*) have it down when they say:

> *"The more information you gather, the better you'll know the right opportunity when you see it—but the more likely you are to have already passed it by.*

> *So what do you do? How do you make an informed decision when the very act of informing it jeopardizes the outcome? It's a cruel situation, bordering on paradox."*

Cruel indeed. You're making a pretty big (some might say even irreversible) decision with limited information. All the while knowing that you can gain more valuable information, but that it would be at the cost of losing equally valuable time.

There's a bunch of complicated math that I'm not qualified to explain, but basically according to the authors, the exact correct balance between looking and leaping is 37%.

Spend the first 37% of your time looking, and once you see something better than the best option you saw during the looking phase, you leap.

I'm not saying you have to follow this exact percentage. Round it up to 40%, or if you don't like math, use a framework that works for you. All I'm saying is that you'll never feel like you have enough information and you'll always feel like you're running out of time. Set a deadline and make a decision when it expires. Be fanatical about sticking to it.

Resist the pressure to extend the timeframe just because you haven't gotten the data you need or to cut the trial-run short because you feel like the world is moving too fast. The right time/information balance exists, figure out what it means for you. It's also important to build trust in your team that there are elements of certainty (i.e. sticking to pre-decided timelines) in this uncertain process.

Organize, organize, organize

The next step is about organizing the data you've collected. Get all your thoughts down systematically in one place, and then make the call on whether you want to Commit, Return or Keep. Here's how I like to do it:

Step 1: Recap the original thinking about why you were going to pivot in the first place. Investors forget. Remind them what the original plan was.

Step 2: Put together a doc summarizing the exploration period. Some basic questions to answer here are — what hypotheses were you testing? What results did you have? What are your conclusions?

Step 3: Build out a decision tree which includes possible outcomes for each category: Commit to pivot, Return to original business, Keep exploring.

Make sure you include all 3 categories because binary decision making sucks. When people are given a "do" or "don't do" classification, folks tend to "do" whatever is proposed. Even when the idea is actually terrible.

Assemble in the editor of your choice, this is usually Google Slides for me since I will likely end up using it in a presentation (to VCs, the board, my team etc). Trim it down as much as you can. I personally like to have everything on just a single slide, it's best if this is factual and objective.

Step 4: It's time to knuckle down now — figuring out which option is the best. This is the hardest bit because it's infinitely subjective. YC's Dalton Caldwell has a helpful formula to evaluate the quality of a pivot idea. You rate each concept from 1-10 on the following parameters and get to a single overall score[26]:

- **How big of an idea it seems to be** — Can it scale in less than 10 years to $100M+ revenue? Can you see it being a publicly traded company in the future?

- **Founder/market fit** — Why are you a the right team to build this idea out? Do you have relevant experience in the space that investors will get?

- **How easy it is to get started** —Are there high barriers to entry (like a big upfront investment or maybe regulatory stuff)? Or is it especially low-friction for you to get into it straight away?

- **Early market feedback from customers** — Do people already want the product/solution now? Will they pay for it?

---

[26] https://www.youtube.com/watch?v=8pNxKX1SUGE

Overall score — Average it out to get a single score out of 10. Rule of thumb: I'd say a 7 or 7.5 is a pretty good score.

Do this for each option in your decision tree. I don't love doing things so mechanically, but putting a number to something keeps you honest in a way no amount of wordy descriptions can. Also you don't have to decide based on these scores, it just gives you a great starting point.

Step 5: Make a final call.

Have a series of conversations around the data you've analyzed, reach a likely decision, sleep on it, and then, make the final call. We'll get into the details of how to do this below.

Talk through the decision with multiple parties

Take this organized information to your investors, to your team, to your board, to your trusted advisors. Tell them the tentative decision your team has landed on. If you've been following the steps in this book, you've already introduced the idea of pivoting to these folks. I have a piece of good news for you: this time, the conversation will probably go down easier.

Now, if you've skipped over to this part, and it's the first time you're going to talk to them about pivoting, this conversation is going to be a toughie. It might be easier if you spread out the meetings over a few weeks, slowly giving them more information, instead of hitting them with everything all at once. Be strong and stay calm.

Either way, keep in mind that you have two main objectives with these conversations:

- **Context + Conclusions:** As a founder you are responsible to the different stakeholders of your company. You have to give them a fair picture of all the biz changes and the rationale behind the decision you're proposing to make.

- **Feedback + Perspectives:** You also have the chance to gather insights from people that have been in the game longer, or have unique industry expertise, or some other value prop. They might come up with a new angle or point out a pitfall you didn't see coming. Make sure you're listening.

Some of these meetings can be intimidating to walk into, especially the ones with your VCs / board. Never forget that everyone in the room wants the same thing — the best outcome for your startup. You also have a lot to gain from their insights, so reframe these conversations in your mind. Go in there with an open-mind and invite feedback, skepticism, and questions. Do your best to answer them. If they ask you something that you don't know yet, be honest about that as well. I'd suggest having all these conversations in the same 48-72 hours to keep things honest. It can be tiring, but there's also something to say about the momentum it generates.

To make this process as effective as possible, a tidbit of advice here: It's best if the company's leadership sits down together to make a tentative decision before getting external stakeholders involved. Get your co-founder(s) and any C-level execs in one room (physically or otherwise!). You can even rope in some key employees for some parts depending on their experience and tolerance for messy conversations. This will probably make the process slower, and it may even cause friction/disagreements. But it'll be worth it in the long run.

How Rupa Health got to conviction

Having conversations with trusted advisors is a huge part of lab testing platform Rupa Health's pivot story. Rupa Health started out as a comprehensive solution for doctors who wanted to set up virtual clinics. Founder Tara Vishwanath saw some growth, but wasn't able to hit PMF.

When her co-founder suggested that they focus on solving just one logistical problem for doctors — ordering labs for their patients — Tara immediately texted a group of doctors with the idea. These doctors were potential customers who had become like an unofficial board of expert advisors for the startup. In an interview with First Round, Tara describes what happened after she hit send[27]:

> *"At that point, we didn't even know what a solution could look like, so we didn't bother pitching a specific product. We focused on pitching the problem. Instantly there was a reaction we had not seen with the virtual clinic platform or the marketplace. For all the other products we built, we were pitching it to the doctors. Now suddenly with one text message, it was the doctors convincing us to build and telling us exactly what was needed. Within minutes, we realized we were on to something."*

Things might not always feel this easy or this clear, but when it does, don't analyze yourself out of it. Go with the momentum.

Consider, then go with your gut

Early in Daniel Kahneman's career, the future Nobel-prize winning economist was tasked with helping the Israeli military hire soldiers. He got interviewers to ask the candidates a set of questions designed to measure traits that might be helpful in combat (like responsibility and sociability). Then, he used an algorithm to weigh the scores input by the interviewers to come up with a final rank for each candidate. The algorithm worked very well. But there was one problem — the interviewers didn't like how mechanical the process had become.

27

https://review.firstround.com/advice-for-the-pre-product-market-fit-days-this-found er's-playbook-for-pivoting-with-purpose#lesson-5-assemble-informal-customer-advis ory-boards-with-short-feedback-loops

Kahneman tried to throw them a bone with another step. After each interview, he asked them to close their eyes, imagine the candidate as a soldier, and put down an overall score. He basically asked them to use their gut. To his own surprise, Kahneman discovered that these scores were even better than the algorithm alone! He had stumbled upon a better way to use intuition[28]:

> "Kahneman believes that the specific questions of his algorithm primed the interviewers to think more analytically about each candidate. This priming ultimately made their intuitive impressions more accurate."

What all this means in the context of pivoting is simple. Be a data-hound at first. Collect as much of it as you can and analyze it. Try to tease out a conclusion or at least a pattern, but also remember to stay open to other possibilities even once you have it.

Now, it's time to use your gut.

Gut intuition has a bad rep because people think it means being rash or biased. Understand that we're talking about something different here — when using your gut to make a decision, review all the data, visualize yourself in the future as a wiser version of yourself, and consider what decision "feels the most right". Be in a relaxed setting, spend unstructured time thinking, sleep on it. Then commit.

And once you commit, don't look back. Research from business professor Laura Huang suggests that gut intuition does not benefit from hindsight rationalization[29]:

---

[28] https://socraticowl.com/post/hire-like-the-israeli-military/
[29] https://hbr.org/2019/10/when-its-ok-to-trust-your-gut-on-a-big-decision

"Once you've decided to rely on your intuition to make a high-risk, high-impact decision, don't try to explain it or justify to others how you arrived at it. If you apply logic and data to gut feel, the more likely you are to put off a decision or make a worse one."

Sometimes we focus too much on logic because we do know we have to explain it—to our team, our investors, our customers, the public. This chapter was all about getting to a decision. In the next chapter, we're going to explore how to explain this decision in a way that softens detractors and creates allies. And it requires a little storytelling flair.

Recap of Chapter 5: Make The Call

Use the Commit-Return-Keep framework

- The story of Intel: memory chips to microprocessors

- Balancing time and information with the 37% rule. (Christian and Griffith's book Algorithms to Live By)

- Recap the original pivot hypothesis, what you did, and what you learned

- YC's Caldwell on evaluating ideas on a scale of 1 to 10

Gather insights from stakeholders

- Talk through the decision with your board members, team, and VCs

- Rupa Health found PMF through an informal board of advisors

Pull the trigger

- Daniel Kahneman's hiring advice to the military (review data, then consider gut)

- Laura Huang: Once you make a gut call, don't try to bring data back in

CHAPTER 6

OWN YOUR STORY

Paint a picture of your company's pivot journey that emphasizes a redeeming core essence while establishing clear reasons for the need to pivot

In a world where everyone's talking about how they hit PMF, one founder took to Linkedin last year to say exactly the opposite.

Ex-Dropbox PM Alexander Embiricos co-founded a startup, hired a team, and announced a splashy fundraise in TechCrunch. But three years later, the story looked very different. In a public post that appeared on thousands of LinkedIn feeds (including my own), Embiricos spilled the truth:[30]

30

https://www.linkedin.com/posts/embirico_remotion-isnt-going-quickly-en ough-to-be-activity-6972969219940302848-WuXI

"Remotion isn't going quickly enough to be a venture scale business. What can we do about it?"

That was a tough conversation between my cofounder Charley Ho and me, as we reflected on our most recent launch."

Embiricos and Ho started building Remotion, a remote working tool, before COVID-19 hit. Their startup was a platform designed to make video calls feel like less of an event and more like an easy way to get something done, without having to go back-and-forth over email. The pandemic-fueled remote work boom gave Remotion a giant push forward. The startup raised $13 million in funding, including a Series A led by Greylock and a Seed Round led by First Round. Embiricos and Ho thought they were off to the races.

Things were great for a while, until they weren't. After the initial growth spurt, the startup's user-churn increased and growth flattened. The team tried everything to onboard more users, but it seemed like people just didn't want Remotion any more. Embiricos and Ho finally decided to face reality head-on. He had a zombie startup on his hands. In September of 2022, Embiricos announced that the startup was going to pivot, taking the bold decision to publically share updates about the process:

"Most critically, we don't think we're learning quickly enough. There are too many "maybes" and "sometimes" in our findings, and individual new features get used by too small a proportion of our audience.

The clear answer: We need to focus. Who should we build for, and what specifically should we solve for them?

We're taking 3 weeks to figure that out. If you're interested in following along or reading our findings, drop a comment below or email me! Happy to share as much of our learnings in this process as I can."

Most people announce that they are pivoting after they've chosen a new direction, when everything is neatly squared away. But choosing to publicly share the messy journey there is rare.

Over the next year, Embiricos got real about Remotion's pivot journey. He shared screenshots from real customer interviews and even pages from their internal notion. The posts went viral and the startup jumped on the momentum, collecting valuable insights in comments like: "We love Remotion on the team (those of us who can use it), however we've stopped using it progressively as half of the team is not on iOs - any changes coming soon on that front? 🫤 "

Embiricos' raw, authentic updates drew people in. It was like a movie. But instead of a scripted Hollywood ending, it was unfolding in realtime on LinkedIn and nobody (not even the director) knew the ending.

The pivot-in-public approach worked for Remotion and after relentlessly course-correcting for almost a year, Embiricos formally announced their new product (and branding) — Multi, a tool for multiplayer collaboration across MacOs.

People hate uncertainty

You might think Embiricos and Ho were crazy for baring it all on LinkedIn. And I'm not saying you need to do it the same way, but there's no denying that you will have to talk about your pivot — to current VCs, team, prospective investors, new hires, customers (old and new).

The problem is that people typically don't react well to this news. All the folks who believed in you before have either withdrawn their support completely or are nervously hanging around to see if it's worth sticking around. People who you're approaching for the first time are hesitant about trusting you.

Don't take my word for it, this is backed up by research. A recent study by management researchers at Cambridge and Northwestern went deep on how to manage user relationships while pivoting[31]. They found that broadly, people either attack the business or start doubting it.

- The attackers are hostile because they feel betrayed by the pivot. They believed you the first time, but they don't want to get burned again. To get them on board, you need to show them the problems your business is facing. Really make them understand why you need to pivot. Shift their focus from betrayal to a wider concern for the business surviving at all.

- The doubters are anxious because they feel confused by the new direction. The uncertainty is scaring them. Here, you need to show them your startup's passion and commitment to follow through on the big picture themes underlying the business. Highlight that the core values / mission of the business are still the same.

In short, you're going to have to do a whole lot of talking about your pivot. It'll be a while before you casually refer to your new thing as the OG biz, without getting into the backstory. All this to say, you need to build a really good story around the pivot.

Pitch like Pixar

Human beings are natural story-tellers. Stories are how we communicate change in an understandable, memorable, and

---

[31] The Art of the Pivot: How New Ventures Manage Identification Relationships with Stakeholders as They Change Direction
https://www.researchgate.net/publication/332213331_The_Art_of_the_Pivot_Ho w_New_Ventures_Manage_Identification_Relationships_with_Stakeholders_as_T hey_Change_Direction

emotionally resonant way. It is one of the most important "technologies" that we've ever invented.

Emma Coats wrote a popular thread about Pixar's storytelling principles after working behind the scenes on Brave[32]. One of them was the idea of the Story Spine, a narrative framework first developed by Kenn Adams[33].

One of the most important elements of the Story Spine is responding to challenge or conflict. Emma Coats writes: "You admire a character for trying more than for their successes." Lucky for us, pivots, by their very nature, always involve a challenge of some kind.

It might feel somewhat contrived to construct a narrative around what was probably an organic process with many unexpected twists and turns. Real life is messier than a story. And yet stories are how we persuade and rally others to our work. It is how we make sense of the world and how others will try to make sense of your company.

As an example, I've broken down Disney's Moana down into this structure:

---

[32] Emma's thread blew up in 2012
https://www.washingtonpost.com/blogs/comic-riffs/post/pixar-tips-brave-artist-emma-coats-shares-her-storytelling-wit-and-wisdom-on-twitter%20followher/2012/06/25/gJQADaxd2V_blog.html

[33] The Story Spine was actually invented by teacher and director Kenn Adams in 1991 https://www.aerogrammestudio.com/2013/06/05/back-to-the-story-spine/

Story Spine	Function	Moana
Beginning (Once upon a time... Everyday they...)	The world of the story is introduced and the main character's routine is established.	Moana is a free-spirited young girl who lives on an island. She longs to sail across the seas, but her father, the island's chief, forbids it as too dangerous.
The Event (Until one day they...)	The routine is broken—either by the main character, or an external force.	The island is in crisis - coconut trees start dying and the fishermen can no longer find fish.
Middle (Because of that x3...)	There are dire consequences for having broken the routine. They struggle to reestablish themselves and we're not sure how things will turn out in the end.	Drawn by the call of the ocean, Moana takes a boat and sails off to save the island. She escapes a giant crab, fights pirates, and teams up with a demi-god.
The Climax (Until finally...)	The main character's prior actions and decisions culminate in a major event, leading to success or failure.	Moana and the demi-god struggle in a fight against a monster. Moana has a crisis of faith before continuing on her mission to save the island.
End (And ever since then...)	The main character experiences the aftermath of their actions, and a new routine is established.	Moana restores the heart of an ancient spirit to heal her island and returns home as a hero.

Now let's apply this idea to your pivot. Here's the same framework but applied to pitching your startup's pivot. While only Remotion / Multi's investors know the exact framing of their pivot, here's how we'd apply the Story Spine to their pivot.

Structure	Function	Remotion / Multi
Beginning (Initial premise, Early success/ wins)	Founders start a company around a specific idea or premise. Early progress looks good.	"We founded Remotion as a workplace platform to make video calls easier. The pandemic gave us a boost, and we raised a seed and Series A off of solid growth numbers."
The Event (Challenge problem)	Founders have a realization while building or an external force causes them to lose conviction in their original direction.	"But over time growth flattened and our users started churning more and more. We iterated extensively and pushed new features but nothing was working."
Middle (Exploration period activities)	Team engages in intensive exploration in order to better understand their customers, the market, and where venture-scale opportunities exist.	"We confronted our struggle head on and publicly admit to the need for a pivot. We surveyed and interviewed many engineers (our target user) and studying existing usage patterns hoping to find something."
The Climax (Realization of new promise)	Through their exploration, the founders identify a new path that can lead to break out success.	"We realized that users often call each other to pair program or work synchronously on a project. So we came up with a way to make real-time collaboration on the same screen dead simple."
End (Early success of new premise)	Early efforts in this new direction show promising results, the future looks bright.	"We've rebranded as Multi, a multiplayer collaboration tool starting on OSX. With 4,000 sign ups on launch week, we know this pivot is the right direction for our business."

I know this is a lot of information. But the bottom line is that you need a really good story for your pivot. This is how you should cover your bases:

- **Initial character is fundamentally good / likable**

  Moana: Free-spirited young island girl with a kind heart.
  Remotion/Multi: Remote working tool that launched in the nick of time.

- **Problems are bad**

  Moana: Island is in danger - coconuts dying and no fish.
  Remotion/Multi: Users churned and growth flattened.

- **Struggle was valiant**

  Moana: Moana encounters grave mental and physical challenges along her journey.
  Remotion/Multi: Founders decide to publicly share their pivot journey, interviewing many users about their collaboration needs.

- **Insight is meaningful / earned**

  Moana: Island is saved because of Moana's chops. Although her love for the island, her community, and the ocean remains the same; her experience transforms her village's approach to exploring the ocean, and her people become voyagers
  Remotion/Multi: Company pivots into a multiplayer collab tool. The essence of the company (to enable collaboration) remains the same but now users have a more powerful way to work together.

Like I said, you're going to use this story many, *many* times — while pitching investors, hiring more people, and marketing to customers. You'll get a lot of practice, and even learn to adapt it to different situations. It matters the most in the first 1-2 years after the pivot; over time, it'll start to fade.

Pivot as PR

A pivot can be a chance to do some major PR for your company. These kinds of stories are interesting, especially when paired with success and momentum on the new product. Figuring out how you tell the story in a way that feels both authentic and interesting to audiences is key.

Case in point: Remotion/Multi. The co-founders spoke about their decision to pivot in public on the Engineering Founders podcast[34]. With their previous product, they got more sign ups by talking about what they were building online. The duo realized that the pivot could be an opportunity to do the same thing, while also staying authentic. Embiricos spoke candidly about their state of mind at the time:

> *"No matter what we find, this is a pretty interesting story... So let's start writing about this. The hope was that people will be interested and when we actually have the product, people will use the product."*

Just like fundraising and original founder myths, the "total truth" is not always what you tell others, and the key is finding a version of the truth that resonates and isn't completely counter to other versions that might eventually emerge. You need to make sure it rings true emotionally, even if that means telling a story that is slightly slanted. Don't lie, but you don't have to give a play-by-play of every detail.

Name change or nah?

Before you go out there and tell your story, you need to decide if you will change your name post-pivot. The big pro of changing your name is the opportunity to start over with a clean state. Nothing holding you back.

---

[34]https://engineering-founders.simplecast.com/episodes/lessons-from-building-in-public-re-discovering-product-market-fit-charley-ho-alexander-embiricos-remotion-3t_zNC03

But this is also the biggest con. You've got to rebuild everything you had worked for from scratch. You lose momentum and name recognition.

As you decide whether your brand needs a makeover, think about these questions:

- Are you pivoting to a completely new area or just leaning into a different niche?

- How much reputation value have you already built up?

- What's the overlap between your current users and target customers post-pivot?

Here are some options to consider:

Just name the product

Before changing your entire company name, try just naming the new product or service. Ford has no less than 5 names for their various SUVs: (Edge, Escape, Explorer, Excursion and Expedition).

Shorten or simplify your name

Often startups insert something concrete into their names to ground what it is and can drop that concreteness if / when it no longer serves them. For instance "Apple Computers" changed its name officially to just "Apple" in 2007 as the iPod reached 50% of total revenues.

Think like a movie producer

From sequels, to spinoffs, to adaptations, many TV shows and movies today are often recycled intellectual property. That's why when you see a film that has brand new IP, they'll often ground it with something you know. In the marketing for the highly underrated original scifi movie *The Creator*, the trailers never failed to mention "From the director of *Rogue One*" to help draw in audiences[35].

That said, if the direction is different enough, it might still make sense to change your name.

If you are still seriously considering that, here is a table that might help your decision based on the two biggest criteria—market reputation and pivot size.

	Big reputation	Small reputation
Major pivot	Worth changing your name. But before you go all-in on a new brand, make sure you're sure. If you aren't ready to fully commit, you can experiment by just shipping under a new product name.	Only change your name if you really want to. Matters less because the market doesn't know you too well anyway. See what works from a time + $$ perspective and roll with it.
Minor pivot	Don't change your name. Leverage your reputation in the new market. The risk here is that people might remember you for your old thing, so make sure your new marketing is super clear.	Don't change your name. Keep at it and build up your reputation with the new product.

Examples of name changes that worked

Our friends at Remotion decided to change their name to Multi in August of 2023[36]. The name "Remotion," Embiricos explained, was

---

[35] *Rogue One's* full title: *Rogue One: A Star Wars Story* is itself is a great example of leveraging a brand people know to introduce something new.
[36] https://www.linkedin.com/posts/embirico_remotion-is-now-multi-a-rebrand-is-a-b ig-activity-7085384151914147840-NIg8

chosen specifically for the startup's original product, the virtual office tool. It indicated emotion and motion, two things the startup originally wanted to help remote teams cultivate. Now that they had pivoted to a collaboration app across Mac OS, in Embiricos' words:

> *"The biggest reason for our rename, though, is that it no longer fits our product."*

Interestingly, Embiricos launched the rebrand just a few weeks before officially announcing the pivot which could have led to confusion among existing customers but might have also built anticipation for the new direction.

Another example of a name change was from Zimride to Lyft. When the company pivoted from long-distance ridesharing to real-time ride hailing, they launched the new service under a whole new brand (and eventually incorporated a separate company to house this biz as well). This name change was an important part of the company's new "friendly" image, including the iconic pink mustaches and the tradition to start rides with a fist-bump.

While we're in the ride-sharing space, we gotta talk about Uber. In 2011, Uber dropped the "cab" from its name. This was prompted by complaints from taxi operators in San Francisco because Uber's model allowed it to tiptoe around the regulations applicable to traditional taxi services. This turned out to be a good thing in the long run because Uber later started offering other services like food delivery and had the option of simply adding a convenient suffix to the parent name—Uber Eats.

When name changes go wrong

Of course, there are lots of times when a name change doesn't work out. Partly why "Definitely change your name" is not in my recommended framework above.

The beloved email marketing platform Converkit had a rare misstep in their own startup journey. In July 2018, ConvertKit announced that it would be changing its name to Seva—which is the concept of "selfless service" in Hinduism and Sikhism[37]. They thought it reflected their ethos more and broaden their brand, but also knew it might be a bit confusing.

"Do you think they'll get it?" they asked each other in their announcement video. Turns out their customers didn't. The community raised questions about whether ConverKit understood the deeply cultural and religious connotations of the word Seva. Reeling from the backlash, they eventually decided to backtrack and rename the business ConvertKit[38].

Handling the old business

There's no shame in shuttering your old business. Truly.

Remember the stat from before—the vast majority of seed funded companies never reach a billion dollar valuation, and most of them are no longer operational 10 years after founding.

You are doing what you can to keep your business alive long enough to get a second chance. But please, wind things down with grace.

You probably have at least a handful of loyal customers — early users who took a chance on you, beta testers that you bothered for feedback, people that you promised new features to. When you decide to shut down, give these kind folk enough time to pull their data, migrate off, and find alternatives.

I dealt with this when my startup Midgame got acquired by Facebook (now Meta). It was bittersweet because we had to shut down and say goodbye to all our customers, a bunch of great people from the

---

[37] https://convertkit.wistia.com/medias/8yxzuto67b
[38] https://convertkit.com/staying-convertkit

gaming community. We'd built close relationships with them over time and it sucked to pull the plug. My co-founder and I penned an article to break the news, where we shared our honest reasons for shuttering. We also included a tiny bit about where each team member was headed in the future. The community responded very warmly:

> *"...All I want to say is huge thank you to the dev team. Thank you for involving me and my friend, it was so much fun to play games, test bot and feel useful at the same time. Love you all, stay safe!"* - Saweczek

Users tend to get upset when they don't have enough time to transition. Just like layoffs, the abrupt nature of the cutoff is the worst. People understand that working with a startup involves change, it's the hurried / curt / unsaid goodbyes that creates animosity.

Whenever possible, I also recommend going one step further — help customers find another service. We did this when we did our pivot from Headlight to Midgame. There was a company called Woven that was also doing take-home technical screens. I reached out to the team and they were thoughtful founders who all had been engineering managers that had collectively hired hundreds of engineers. A great example of a story and a set of assets (in the form of earned credibility) that made them a better team to go after this market. When we pivoted, we arranged a "PR acquisition" that allowed Woven to announce they had acquired Headlight. This allowed them to do some press and for us to explain to our customers that they could continue getting their services through us.

You want to make sure you leave things on the best note possible, people will remember the way you treated them the next time you launch a product/service. The first step here is finding a reliable service provider who could take your place. Ideally this should be someone who you've worked with before and can confidently endorse. Try to make the transition as seamless as possible, and then proactively check-in to ask if things are going well.

Moving forward after a pivot

Congratulations, you pulled off a nearly impossible task.

Deciding to explore a potential pivot, finding a new opportunity that's a fit with your strengths and experience and pulling the trigger on a major shift is hard as hell. Many entrepreneurs have attempted this feat and failed.

But you have a long way ahead of you. You need to build momentum in a new direction, change your processes to suit your new business, recruit new talent, raise more funding, and push through the inevitable ups and downs (mostly downs if we're being real) of entrepreneurship.

Part of the challenge is seeing yourself in a new light. You're no longer Jane Doe, founder of Acme Startup, the #1 Product Hunt of the day. Now your startup has a new name and people aren't totally sure what it does.

That's not such a great feeling. You may find yourself questioning whether this pivot was the right decision. You might look back fondly at your old business, which had users, and a working product, and wonder what foolishness brought you here.

But don't forget. You had a zombie company, an undead business that was never going to get you to the promised land. You still might never get there, but what you have done is bought yourself a chance.

A chance to build something that can scale, that can hit hypergrowth, and change the world in a shorter timeframe than is reasonably possible in any other permutation.

You are a startup founder reborn. And if you are to face the adversity ahead, you need to invest in yourself. You've forded the river in the Oregon Trail of startup success. Snakebites, dysentery, and starvation are still looming threats—and yet, you are still alive. You haven't wasted your shot.

With this chapter, we wrap up the Pivot Blueprint, which covered the most essential and unique ideas from *The Path to Pivot*.

Next up is Section II: the Pivot Toolkit. Here we'll go over some of the tactics and systems you can use to execute your pivot with greater impact while avoiding common pitfalls. If you are a faithful student of entrepreneurship, you may find some of the chapters may feel familiar. Though I promise you'll learn something.

If you're a young gun and this is your first company, consider the reader a crash course in building a venture-backed startup.

Feel free to wrap up this book at this point if you feel equipped to go ahead and make your magic happen. Come back when you have new questions or need help. Alternatively, you can keep reading ahead now.

Recap of Chapter 6: Own Your Story

Weave a compelling story around your pivot to win over investors, customers, and employees

- Pitch like Pixar. Use the viral 5-part Story Spine
- Leverage your pivot for PR (like Remotion)

Revamp your identity

- Framework to decide if you should change your name
- From Convertkit to Seva and back - story of a name change gone wrong
- Help customers transition if you can't support them

"They always say time changes things, but you actually have to change them yourself." — Andy Warhol

Section II

PIVOT TOOLKIT

In this section, you will find practical strategies and actionable steps to execute your pivot with confidence and impact. We cover getting stakeholder buy-in, managing the finances, navigating the idea maze, and staying resilient.

Feel free to skip around, read out of order, and focus on the chapters that are most relevant to your situation.

CHAPTER 7

ALIGN INCREMENTALLY

Making the decision to pivot can be contentious and not everyone will be on board—follow the right sequence to ensure full commitment

There's an old joke about a dead cat that's the perfect metaphor for how you align your team around a pivot[39].

A man left his cat with his brother while he went on vacation for a week.

When he came back, he called his brother to see when he could pick the cat up. The brother hesitated, then said, "I'm so sorry, but while you were away, the cat died."

---

[39] I don't know if there's an official source but I found this version from (very aptly) an IT leadership blog:
https://thenamiracleoccurs.wordpress.com/2012/07/17/what-to-do-when-grandmas -on-the-roof/

The man was very upset and yelled, "You know, you could have broken the news to me better than that. When I called today, you could have said he was on the roof and wouldn't come down.

Then when I called the next day, you could have said that he had fallen off and the vet was working on patching him up.

Then when I called the third day, you could have said he had passed away."

The brother thought about it and apologized.

"So how's Mom?" asked the man.

"Well, she's on the roof and won't come down."

The lesson: If you can, break bad news sequentially, so your audience can better anticipate and process it. And of course, do so in a way that is contextually relevant. That's what we do with the Align-Explore-Commit process.

If you aren't playing as a team, you've already lost the game

Alright, you've finally decided that your startup needs to pivot. It may have taken some soul-searching, but you've acknowledged that something just isn't working. It's a bitter pill to swallow. I'm going to be straight with you, it's about to get harder. Because you have to break Bad News and no one wants to do that. But you have to do it, and do it sooner rather than later.

To see why it works, let's look at the opposite approach.

Let's say you decide you have the runway and energy/motivation to explore and pursue a pivot. All the founders and company leaders are aligned, so you immediately tell everyone to stop what they're doing and start working on the new thing.

How are they going to feel? Will they deliver their best work under these conditions?

As Jessica Livingston of YC wrote in *Founders at Work*:

> *"People like the idea of innovation in the abstract, but when you present them with any specific innovation, they tend to reject it because it doesn't fit with what they already know."*

This holds true even for the people doing the innovating!

Changing minds takes time and evidence

If you've done your job well as a founder, you've shielded your stakeholders from the doubts you've had.

- The CTO you found on YC's co-founder matching platform, who worked countless weekends to ship new features ready every Monday.
- The first three hires who took big pay cuts to join you pre-revenue.
- The VCs who you had coffee with and talked a big game about your upwards trajectory.

They don't know about all the difficult feelings you've worked through over the last few chapters, this sense that you're a quitter, that you don't want to give up what you have, and that loss of identity. It's still a process they gotta experience their own version of.

Cofounders + Employees: Work is an important part of people's lives and working at a startup is inherently more risky than a corporate job. So while your team might be more capable of handling a routine shift like losing a major client or having to deal with a major tech issue with the app, walking in one day and being told that everything is changing can really be jarring.

Investors: It's true that VC's are more used to seeing companies pivot, and you're only one of the companies in their portfolio, so your big shift doesn't affect them as much as your employees. Still, they are a key stakeholder who might give you more funding or intro you to new investors for your pivot, so you don't want them to feel blindsided about the pivot.

This is especially possible if you've been silent for a long time or if the last time you talked you were very gung ho about the current business. They talk about your business (as your investor) to other people and hate to feel foolish in not knowing things have totally changed.

So what do you do?

As the IT leadership blog which gave us our chapter-opening dead cat joke explains, the only thing worse than delivering Bad News to our stakeholders, whether the Non-IT folks, investors, or anyone else, is letting them down late:

> *"The natural inclination on our part, nice people that we are, is to spare them their unhappiness by delaying the Bad News as long as possible. After all, maybe something good will happen that'll make it all come out right – maybe a Miracle Will Occur, and the Bad Thing will be avoided. So we put off delivering the disappointing news as late as possible.*
>
> *We shouldn't do this. The one thing that's worse than getting bad news is getting bad news at the last minute."*

Getting the decision sorted

Few company decisions stir up as much controversy as a potential pivot. Everyone's got an opinion. But you can only choose one path forward.

Pivots can divide teams. Yet, it's also one of those times where a collective commitment to forge ahead is invaluable. As captain of the ship, it's your responsibility to get everybody aligned.

The order in which you have these conversations matters. Start by going to your co-founder and senior company leaders. Sit them down, offer them a warm beverage, and have an honest conversation with them. Don't hold back. These are your closest strategic allies — you might not agree on everything, but you share a broader vision of a successful company.

Once you've settled on a strong message that leadership can stand behind, it's time to hit up your investors. This conversation might feel especially intimidating. They trusted you with real money to execute a specific idea — and now you're coming back to tell them that it didn't work out as planned. You might feel an odd sense of guilt gnawing away at you, like you've failed them.

Pause. Remember that they've seen this happen before; over, and over, and over again. If you're transparent about what's happening, their experienced insights will be instrumental in shaping your pivot strategy.

When alignment goes right

I'll say it again because it's important — it's tempting to hide the true state of affairs from your VCs because you're scared or because you want to save face; don't. Your investors might surprise you.

Take the story of the mobile-linking unicorn Branch for example. The founding team's original product was actually a photobook app targeted at mothers and teenagers. After they pivoted into deep linking, their angel investors Pear VC revealed they had never been too keen on the idea of the photo app. They had invested because they believed in the team, not the idea.

After your investors comes the rest of your team. Convene a company-wide meeting and communicate the pivot to them as an informed decision backed by leadership. Keep in mind that this is an announcement, not a vote. A clear tone will foster confidence and swift action.

A roadmap for the talk

At the end of the day, these are tough conversations to have. I've found that it always helps to go in with a framework. Something that looks a little like this:

- Start with explaining what convinced you to reconsider the company's direction. Get into the details and call out the problem — whether its declining growth, stagnant revenue figures, or changing market dynamics.

- Paint a picture of the company over the next few months if everything were to stay the same. If you're at the point of diminishing returns, this will clearly show that you'll have to work harder and harder to achieve even small growth upticks.

- When you're chatting with company leadership, remind them that your investors and employees are not dumb. They know that things aren't going very well. It's better to change direction before they start leaving a sinking ship.

- Showcase opportunities that show promise and propose the idea of pivoting to explore these avenues. Talk about the potential benefits of tapping into new markets or capitalizing on emerging trends.

- Reiterate what you still believe in — the product, the market, the team, or the need to solve a specific problem. Explain that while the current path is not working, a pivot is needed to honor these beliefs.

If key stakeholders are aligned on the decision to pivot, that's great. Head on over to the next chapter to work out how much money you need.

Knowing when to fold

What if after all that, you're still facing a lot of pushback? You've had the conversation with your cofounder (or other key stakeholder) over and over *and over again* and for some reason you just can't see eye to eye on the pivot. Don't worry, I got you.

They're feeling uncomfortable because ditching a idea has such a bad rap in Silicon Valley[40] and they're caught up in the cognitive biases we talked about in Chapter 1.

In this situation, I suggest taking the advice of former pro poker player Annie Duke and start by agreeing with your cofounder.

You're right, we can't just give up on this idea yet!

That might be the opposite of what you think but just bear with me. In her book *Quit*, Duke asked legendary VC and self-described quitting coach Ron Conway what he does when founders don't agree with his advice to quit[41]. Conway had a helpful 4-step framework:

Step 1: Start by putting forth the idea that the founder should consider quitting the venture.

Step 2: When the founder resists, step back and agree—that yes, they could salvage the situation.

Step 3: Establish precise criteria for what a successful outcome would look like—use these as decisive factors for future assessment.

Step 4: Agree on a future date to review the situation, with the understanding that if the benchmarks for success have not been met, a serious conversation about quitting will be necessary.

40

https://review.firstround.com/grit-or-quit-tactical-advice-for-founders-facing-tough-company-building-decisions#call-it-by-its-name-why-a-pivot-is-a-quit

[41] Conway prides himself on getting founders to understand when it's time to quit and helping them through it

This strategy is all about incepting a threatening idea in an incremental fashion, and guarding against the biases we talked about earlier. Duke has a great piece of advice about the kill criteria in Step 3[42]:

> *"A simple way to develop the kill criteria is with 'states and dates'. If by (date), I have/haven't (reached a particular state), I'll quit."*

That's not to say that it's all frameworks and formulas. It's important to get real with your cofounder/other key stakeholder. Talk about the tough emotions you're feeling. Tell them you don't want to waste your life trying to make an idea you've lost faith in work. Paint them a picture of what it's been like for you.

Regroup with them at the end of the test period. Touch base on the benchmark criteria you had agreed upon. If you still can't agree on a way forward for the company, it's probably time to part ways. You don't agree on a direction anymore, there's no shame in stepping away. You can refocus your time, energy, and resources on figuring out the most graceful and efficient way to do that.

※

Once you've reached alignment on moving forward with the pivot, your next big task is making sure you have enough resources to do so. Let's get into the money conversation in this next chapter.

42

https://review.firstround.com/grit-or-quit-tactical-advice-for-founders-facing-tough-company-building-decisions#call-it-by-its-name-why-a-pivot-is-a-quit

Chapter 7 Recap: Align Incrementally

It's important to break news of the pivot sequentially

- The "cat on the roof" method
- Co-founders first, investors next, employees last

Roadmap for the talk

- Sooner > Later. "The one thing that's worse than getting bad news is getting bad news at the last minute"
- Frame the convo: the status quo is unsustainable, people aren't dumb, what do you believe in?

How to deal with the pushback, if any

- Conway's quit criteria—if not now, then when do we revisit this decision?
- "States and dates" can help make the next convo more unbiased

CHAPTER 8

EXTEND YOUR RUNWAY

Pivots aren't cheap—give yourself enough time to land your pivot by cutting expenses deeply and getting creative about acquiring additional cash

What kind of founder turns down millions of dollars in a Series A check for two bridge loans totalling $400k?

Meet Ooshma Garg, the founder of meal kit company Gobble.

From 2011 to 2014, Gobble evolved through a series of iterations. Starting off with a vision of bringing authentic, home cooked meals right to your doorstep, the company found itself in the midst of a unique problem: scaling authenticity. Imagine asking an Italian mom to suddenly prepare ten times her signature lasagna. It's a recipe for logistical chaos and diminishing the heart of the dish.

In their quest for profitability, Gobble pivoted towards a seemingly lucrative avenue—catering. As cash flowed in and big names like Box and Pinterest became their clients, a tricky fork in the road emerged.

When the time to raise a Series A round arrived, investors were more interested in the enterprise side, which had grown to constitute 75% of their revenue. But was that Gobble's true identity?

Ooshma's vision was to help everyday consumers enjoy home cooked meals at home. So instead of raising millions and going all in on catering, she raises two $200k bridge round, winds down the catering business, and retrenches to rebuild her company.[43]

The story ends well: after several more iterations, Gobble was sold in a mouth-watering (sorry) nine-figure deal to Intelligent Foods in 2022.

Let's get down to brass tacks—money. To successfully pivot your startup, you've got to have the funds and the time. You're essentially hitting the reset button, taking a step back to leap forward towards a brighter, more lucrative future. So how do you extend that all-important runway? You've got two levers to pull: reduce your burn rate and boost your bank balance.

Reducing expenses

To extend your runway, you need to cut down on costs. Your biggest culprit is headcount, and for in-person teams, office space. We'll focus on software-based ventures, hardware-driven businesses may have different considerations.

Layoffs

Layoffs are a hard pill to swallow—I get it. However, you have to face the facts: it's likely your biggest expense by far.

You've got two timing options for this grim task—do it right at the outset or wait a bit. The beauty of the "Align-Explore-Commit"

[43] https://www.justgogrind.com/p/ooshma-garg

framework is that it provides your team with a heads-up that things are shifting. It allows both underperformers and stars to show their true colors in a crisis.

If you already know that you're completely pivoting away from a specific line of business, you may want to sever ties with experts or contractors in that area immediately. But if you're still weighing options, consider a one-month "pivot period" before making any rash staffing decisions.

Pay cuts & deferrals

Another lever you can pull either instead or in tandem if you genuinely value every member of your team and believe they're all essential, is to do pay cuts.

The blow can be softened with deferrals—where you promise higher compensation when new funding is secured and offering additional equity, which aligns employees with the business's success. Not everyone will accept this arrangement, but those who do demonstrate their commitment and belief in the company's potential.

If you do pull off a pay deferral, know that this is a loan that will come due. You will lose pretty much all credibility and trust from your team (and tank the company) if you don't make good on your promise eventually. Plus you'll give all founders a bad name. Please don't do that.

Expense optimization

Look for ways to reduce expenses across the board—get rid of tools you aren't using, look out for annual renewals that might be coming up. A lot of license products will offer a discount if you attempt to discontinue, but not always so be careful.

Don't forget to negotiate payment terms with your vendors. You might be surprised how willing they are to extend from net 30 to net 60 day terms.

A lot of startup vendors understand the risks associated with working with early stage companies and may be flexible. Engage with your law firm, accountant, and other service providers to explore potential payment deferrals or alternative arrangements.

When Notion retrenched

One of the most famous examples of a startup team cutting to the bone during a pivot is Notion. Three years into their journey, the startup had raised $2 million from angel investors and hired 4 employees with nothing but 2 dud product launches to show for it. They were burning through money, fast. Founders Ivan Zhao and Simon Last took the call to start afresh.

They laid off their young team and moved from SF to Kyoto (which was half as expensive as Silicon Valley). Zhao and Last lived on one floor of a two-storey house so small that only a traditional Shoji screen separated their bedrooms. In an interview with Sequoia, Zhao recalls working 18-hour days: "We were just, code, code, code. Then, 'Hey, let's go out for food.' Then, we go eat, go back to work, and do it again."[44]

Increasing Cash Flow: Funding Your Pivot

To extend your runway, you must explore avenues to bring in additional capital. Here are two practical methods:

Upsell existing customers

If you're already generating some revenue, your current customer base might offer some additional runway. If it's not too hard to keep operating your existing product, you can consider getting some quick cash for a longer commitment.

---

[44] https://www.sequoiacap.com/article/notion-spotlight/

Offer them a discounted annual plan if they pay ahead of time. If you're already doing that, try bumping the discount for a limited time to get folks off the fence.

Like the pay deferral, this is again a borrowing against the trust you've built. Don't take money for services you can't deliver. People remember, and you don't want to make enemies who are going to badmouth your new direction.

Get paid to build

A great way to find the right product might be to change the conversation you have with customers, in a "consultingish" manner as Paul Graham suggests in his article "The Pinch"[45].

"There is a long slippery slope from making products to pure consulting, and you don't have to go far down it before you start to offer something really attractive to customers. Although your product may not be very appealing yet, if you're a startup your programmers will often be way better than the ones your customers have. Or you may have expertise in some new field they don't understand. So if you change your sales conversations just a little from "do you want to buy our product?" to "what do you need that you'd pay a lot for?" you may find it's suddenly a lot easier to extract money from customers."

This system can also help you figure out where you want to pivot, but watch out: you may come to find that the people who would happily pay you a lot to do the work for them might not want to pay less for a software tool that does 85% of the work but leaves the last 15% on them.

Acquiring new clients

Once you have clarified your pivot and established a landing page, reach out to potential customers in your new target market. Offer them early access or pre-orders, collecting payments in advance.

---

[45] http://www.paulgraham.com/pinch.html

I know of a YC-backed healthtech company that used this strategy when going through a pivot. Despite making low six figures via enterprise contracts, they felt their overall market size was limited. They knew their new product direction, still in healthcare but focused on a different problem, was a hit when they were able to land a $250k contract from a single client with just a Figma prototype. Selling ahead of working software takes guts but it's absolutely do-able. Just make sure you deliver in the end.

Presales not only generate cash flow for the business but also validates market demand and is a great signal that you're on the right track. We'll talk more about that in Chapter 9.

Raising capital

If you have a close relationship with existing investors who support your pivot, you may consider approaching them for a bridge round or an extension. This may dilute your equity, but it can provide a short-term financial boost. Alternatively, seeking new investors might be necessary, but it is often more feasible after you have made progress with your pivot.

As of the time of this edition's publication in early 2024, the time of easy money may be temporarily rolled back, but this strategy will always be an option for a truly promising venture-backable company with a reputable founder. As market conditions change, it may become available to you as well.

Giving it back and starting over

I know this chapter has been about ways to get more money, but paradoxically, sometimes you gotta give the money back.

When Nuvocargo founder Chuggani decided to pivot from a recruiting startup to a freight logistics platform, he did something everyone advised him against—he offered his investors their money back because he was pivoting. YC partner Jared Friedman remembers it years later[46]:

> *"If founders decide to pivot their company, it's uncommon for them to offer the money back to investors and there's certainly no legal obligation. I had a lot of respect for him doing something that could have killed his company just because he felt that it was the right thing to do."*

Most of the investors (Friedman included) decided to stay invested in Chuggani. For the few that did take him up on the offer, Chuggani was able to convince the remaining investors to put in additional sums that made up for the lost investment.

If any of your investors take their money back, my two cents is that you're actually better off. I don't mean that in a bitter sour-grapes way. I meant that you don't want an investor who doesn't really believe in the biz in your cap table.

If anything, use it as motivation. Make this the biggest mistake of their investing career.

Why starting over on a pivot sometimes makes sense

Returning to our friend at Union Square, Fred Wilson also says sometimes it's better to quit and start over, or do something else, that startups are not "indentured servitude".

He is not a fan of the hard pivot that goes in a completely different direction as it can make things complicated for investors and the founders themselves. As much as money is a key concern for an investor, the hassle of dealing with a company that the VC no longer has interest in is also a factor.

46

https://www.forbes.com/sites/alexandrawilson1/2020/03/02/his-first-idea-failed-this-26-year-old-immigrant-convinced-his-investors-to-back-his-next-idea-anyways/?sh=5447b3861711

As he writes in Inc Magazine[47]:

> *"My view is if you've failed, accept it, announce it, and deal with it. Shut the business down, give back the cash, and rip up the cap table. Then do whatever you want to do next. If it is another startup, do it from scratch and keep as much of it as you can. If it is something else, well then do that too."*

Sam Lessin of Slow Ventures makes a similar point, warning that founders can lose control of their venture more easily when 30% of the cap table is "dead" from previous investor ownership[48].

> *"The greatest companies are built by founding teams that are deeply in control, and even with very supportive capital, when founders lose the ability to outvote everyone else I believe they make worse decisions out of fear for their safety and position."*

I definitely ran into this issue when Headlight was pivoting to Midgame. We took 3 rounds of fairly dilutive pre-seed / accelerator funding and when COVID hit and we were faced with the prospect of needing to shift gears again, the prospect of having so much of our cap table used up made the decision to pursue some kind of exit from the business[49].

The bottom line on the bottom line

Chances are you'll have to make some tough cuts. It's a painful but necessary step for the greater good of your venture. If you've never laid off a significant portion of your team, it can feel incredibly scary.

But the truth is, staff cuts are often incredibly effective for companies and you'll discover that outside of the real sadness of the human cost, absolute performance can grow.

---

[47]https://www.inc.com/fred-wilson/pivot-business-failure-start-over-investors.html

[48] https://www.theinformation.com/articles/the-hidden-costs-of-pivoting

[49] We attempted to recapitalized the business but the proposal fell apart last minute

In the Lenny Podcast, executive coach Matt Mochary described how every one of his coaching clients had to do layoffs during` the COVID-19 pandemic, cutting between 5% and 40% of their staff[50]. And what they found across the board was actually a lift in performance. As one CEO told him within 60 days after the layoff:

> *"It's insane. I don't know how this happened, but the company's now operating better. I'm not talking on a relative scale, I'm talking on an absolute scale. We're putting out more features, more code. Our NPS is up, [every] department is performing better. The only answer for it was we've got less people, so this coordination issue is reduced."*

It's a bitter pill, but think of it this way—it's better to lose some parts than for the whole machine to fall apart.

Extending your runway during a startup pivot requires a diligent approach to managing expenses and acquiring funds. While reducing burn through layoffs and expense optimization is necessary, it is equally important to explore methods to increase cash flow, such as leveraging existing customers and acquiring new clients. Raising capital from existing or new investors can provide a crucial financial cushion. Remember, making tough decisions and prioritizing the sustainability of your business is essential for long-term success.

Now that we've thoroughly preserved all the financial resources we have on hand, it's time to use that additional time to get out into the world, particularly, the lives of your customers.

---

[50]https://www.lennyspodcast.com/how-to-fire-people-with-grace-work-through-fear-and-nurture-innovation-matt-mochary-ceo-coach/

Recap of Chapter 8: Extend Your Runway

Reducing burn is a must

- The Notion founders decamped to Japan to cut costs
- Bitter pills - layoffs, pay cuts suck but have unexpected upsides
- Get creative about optimizing other expenses

Yes, you can increase your cash flow even now

- Upsell existing clients on annual plans
- Get paid to build with PG's "consultingish" strategy
- Presell to get new clients and validate market
- Raise bridge capital if terms aren't terrible sense

Offer to give it all back and start over

- If that sounds wrong, remember Chhugani at Nuvocargo
- Wilson (Union Square Ventures) and Lessin (Slow Ventures) think so too
- Branch's founders were backed again after dropping the photobook idea

CHAPTER 9

GET OUT OF THE BUILDING

The answers to your pivot are out in the world in the lives of your customers—time spent understanding their problems is never wasted

This chapter is all about getting to know your customers, which requires going above and beyond to spend time in their shoes. The team at Beacons took an extreme but insightful approach: they had a customer live with them rent-free to learn from her.

The Stanford PhD grads first built a hardware lock but pivoted after talking to dozens of potential customers and realizing they felt uncomfortable with the constant face monitoring the device required. By the time they started YC they were exploring other ideas. Their CEO Neal then met Shopify seller Xue at a friend's wedding, who convinced him to use their machine learning for influencer marketing tools.

Xue had quit her finance job at JP Morgan to focus on her e-commerce business but struggled with the tedious process of finding and contacting influencers to pay for promotion. Learning their lesson from their previous idea, Beacons proposed an unusual solution to get more customer feedback: "We told her we'd build the product she wanted if she'd come live with us and give us feedback on it," Zhang said. To Xue's credit (Zhang calls her "one of a kind" for this move), she agreed.

This immersive customer insight, combined with attending an LA creator conference, showed the bigger opportunity was helping influencers monetize brand interest. Beacons launched with Cameo-style buy-a-video feature, boasting a paltry $20 in revenue from a single user at Demo Day, but their instincts proved right, leading to hundreds of thousands of users and a $6M seed round led by a16z less than two years later.

The Beacons story demonstrates the power of stepping into customers' shoes. Their extreme but clever approach offers an unforgettable lesson in learning deeply from the very users you aim to serve.

When you make the decision to pivot, you need to adopt an explorer mindset. You're no longer on the well-trodden path; you're hacking your way through the startup jungle, searching for that elusive treasure—a product-market fit that clicks.

The Explorer mindset

Being an explorer in the startup world is a lot like being one in the wild. You need to be curious, observant, and optimistic. You need to listen between the lines and believe that there are still untapped opportunities out there. Jeff Bezos once said that customers are "beautifully, wonderfully dissatisfied," even when they report being happy. This dissatisfaction is your goldmine. It's your job to dig deeper and find out what's really bothering them.

Talking to customers is the heart of Exploration

This is where the rubber meets the road. You've got to talk to customers. For some, this is the hardest part because it takes you out of your comfort zone of dreaming and building. But if you've built a business before, you know the drill. You've got to get out there and engage with the people who matter most—your users or potential users.

Uncover dissatisfaction. There are many advantages to a customer-centric approach, but here's the big one:

> *Customers are always beautifully, wonderfully dissatisfied, even when they report being happy and business is great. Even when they don't yet know it, customers want something better, and your desire to delight customers will drive you to invent on their behalf. — Jeff Bezos (2016 Shareholder Letter)*

Most founders suffer more from failing to find demand than failing to meet that demand. Which means if your company is stuck, it usually hasn't identified the right problem to solve. Your solution is to talk to people and find out what they're dissatisfied with.

Bricks and hair on fire problems

A common thing you'll hear when working in startups is that founders should look for "hair on fire problems"—meaning something so severe that your customers are willing to try anything to get the job done. Y Combinator's Michael Siebel elaborates on this[51]:

> *If your friend was standing next to you and their hair was on fire, that fire would be the only thing they really cared about in this*

---

[51] https://www.michaelseibel.com/blog/the-real-product-market-fit

world. It wouldn't matter if they were hungry, just suffered a bad breakup, or were running late to a meeting—they'd prioritize putting the fire out.

If you handed them a brick they would still grab it and try to hit themselves on the head to put out the fire. You need to find problems so dire that users are willing to try half-baked, v1, imperfect solutions.

The Mom Test and beyond

There is a lot to say about talking to people in order to uncover problems and I highly recommend you read *The Mom Test* by Rob Fitz. Published in 2013, it has over 1,000+ 5 star reviews because it's simple, clear, and explains how to actually validate our business concepts. Here are the most important points from the book:

1. Make sure to talk to people in your target market to confirm whether you are solving a real problem.

2. Focus on getting facts about their life, describing their problem in their own words, how much it matters and what they've tried to do to solve it.

3. If you do talk about potential solutions or anything you've built, ignore compliments and vague statements of support — most people (especially friends & family) want to be nice and not crush your dreams.

4. Focus on getting true signal: are people willing to introduce you to others, are interested in follow-up meetings, or want to part with their money. Everything else is a "no" / "not interested".

Note: Another reason why interviewing users is so important when pivoting is that if you can't get a hold of enough people in your target audience just to have a dozen conversations, how the hell are you going to reach them when you have a product to offer?

Reading between the lines

Paul Graham, the co-founder of Y Combinator, advises founders to explain what they've learned from their users. This is crucial. Learning means updating your model of the world, discovering something new, something you were wrong about. That's what great explorers do—they venture into the unknown and return with discoveries.

Segment co-founders had 2 failed products (a classroom lecture tool and a customer data analytics product) and six months of runway left when they decided to give it one last shot. They went with an internal tool they had built to organize analytics data from different sources. It saw traction immediately when they launched it on Github. When the founders decided to go all-in on the idea, they decided to do things differently this time round. As they recall[52]:

> *"The most valuable thing we ever did at this stage was enable livechat via Olark. It gave us an unparalleled window into what users were asking for. Unlike our earlier ideas, we were now getting constant feedback on the product. People would start using the product and then just kept asking us for more."*

The art of the interview

When you're talking to customers, you're not just gathering data; you're also building a relationship. Ask open-ended questions that get to the heart of their experiences and frustrations. For example, if you're exploring the pet toy market, you might ask questions like, "What bothers you most about your pet's current toy?" or "Do you ever worry that your pet is bored when you're not home?" Each question is a stepping stone to deeper understanding.

---

[52] https://segment.com/blog/show-hn-to-series-d/

Reading Between the Lines

Paul Graham once said that his #1 piece of advice for founders trying to get into Y Combinator is to explain what you've learned from your users. This implies that:

A. you have users and
B. you've learned things from them.

To learn something is to update your model of the world. And that means discovering something new, something you were wrong about. That's what great explorers do—they go into the unknown and return with discoveries.

Example interview questions for a pet company

Let's say you're interested in exploring the pet toy market. Here are some questions you might ask:

Background Information:

1. What was the last pet toy you got?
2. Did you buy it or did someone give it to you?
3. When did you last use it?

Evaluating Pet Enjoyment:

4. Did your pet like it? How can you tell?
5. Is this your pet's favorite toy? If not, what is?
6. Why that toy?
7. What are other things your pet enjoys doing besides playing with toys?

Toy Features and Preferences:

8. Is this a toy your pet can play with on its own or do you need to use it as well?
9. What made you choose that toy over the others?
10. What bothers you most about it?

11. What would make it more convenient for you?

Pet Relationship and Enrichment:

12. What is your favorite memory with your pet?
13. Do you have any keepsakes of your pet?
14. Why do you buy toys for your pet?
15. How does it make you feel when your pet really likes a toy?

Shopping Behavior:

16. Where do you discover new toys your pet might like?
17. How do you evaluate whether your pet will like a toy before you buy it?
18. How do you evaluate whether a toy is "worth" buying?

You can direct these questions towards your existing product if you have enough users, or focus on a new product or problem area.

Ask better questions with AI: With the emergence of generally knowledgeable LLM's, you can now ask ChatGPT to generate a day or week in the life of your target customer, and use that to refine the questions you ask so you can really hone in on a pain point they need solving.

The goal of these customer interviews is to identify something new you didn't already know about your users and their life, something you could use to solve their problems better and build a big company around. Each one of these questions could take 3-5 minutes to answer so you see how follow up conversations are often necessary if you have an owner with a lot of opinions

How many people should you talk to?

Research trips challenge our preconceived notions and keep clichés at bay. They fuel inspiration. They are, I believe, what keeps us creating rather than copying. —Ed Catmull, Creativity Inc.

Founders will be tempted to do 5–7 conversations and think it's enough to get a signal on the problem. And if you were simply doing a user test on a single screen or workflow in an app, that *might* be true.

If talking to 5-7 people yields no visible opportunity, it may be fine to move on. But if you find something interesting it's worth expanding out to 20-30 or more conversations—especially if you are exploring a large market or broad problem area. While narrowing down is important as we'll discuss in the next chapter, it's also dangerous to latch onto *too* specific of a problem or use case.

One startup I know had a Google Drive with 100 potential customers they explored during their pivot. Rather than seeing this research work as tedious, understand that it will feed your creative process, just as Pixar's Ed Catmull explains. Remember that often you can't cover everything in your initial call and might need to do a follow up to really understand their world.

The importance of demographics and documentation

Talking to customers should be a team effort, not a task delegated to a single "customer person". Your entire leadership team should be talking to users or potential users, because everyone's perspective is different, what they will probe on or ask further questions on is different.

Make sure to capture some basic demographic information from each conversation—gender, age, location, relevant details to the problem area like number of pets, age of pets, urban/suburban dweller, etc.

With tools like Zoom and Descript, you can easily record and transcribe your conversations for future reference, but don't use that to skip writing notes—by forcing yourself to record what is important, you will learn more.

Remember, your efforts in finding people to talk to will end up being similar to how you market / sell to these people later on, so the work is rarely wasted if you decide to go in this direction..

When Gobble founder Ooshma Garg said she was going to understand what her customers wanted, she meant business. At the time, the startup was a dinner subscription service for busy people who wanted wholesome meals. Ooshma and her team visited their customer's homes and watched them eat these dinners to get real insights. In a YC conference[53] she says:

"It was kind of awkward because we were standing in the corner of their house, while they were cooking and talking to their spouse or their kids were on their legs. Then, they were eating and we were watching them eat. That's what you have to do!"

It was all worth it in the end because the team picked up on something that changed their business model—people didn't like microwaving food for their loved ones. People wanted to cook, they just didn't have time to. Gobble pivoted from doing dinners to serving up meal kits that people could put together themselves.

⁂

Pivoting is not just a shift in product direction; it's a journey of discovery. It's about being an explorer, armed with curiosity and guided by the voices of your customers. It's about asking the right questions and being willing to listen, really listen, to the answers. It's about being open to surprises and learning from them. And most

---

[53] https://www.ycombinator.com/blog/ooshma-garg-female-founders-video/

importantly, it's about being willing to venture into the unknown, to be wrong, and to come back wiser.

Once you find an opportunity, problem, or market segment you're excited about, your brain might be filled with big visions of a world changing product. Hold onto that dream.

But at the same time, carve out space in your mind to get super specific. Just like how you gotta fight off all Koopas before you face Bowser in Super Mario, you gotta overachieve in a tiny sliver of this opportunity and earn the right to play bigger.

Recap of Chapter 9: Get Out Of The Building

The Explorer mindset for developing your market

- Look for "beautifully, wonderfully" dissatisfied customers
- Don't be afraid to get extreme with customer feedback techniques
- Gobble - Founder Garg watched customers make dinners
- Beacon - had their first customer live with them!

Reading between the lines - discover problems with better questions

- Read The Mom Test for practical interview strategies
- Focus on the right signals. Praise << Willingness to spend money or take another meeting
- PGs #1 advice for YC-hopefuls: "What have you learned from users already?"
- Don't delegate this to the "customer person"—every leader needs to get their nose into your customer's business
- Repeat over and over and over and over

CHAPTER 10

NARROW DOWN

Choose a specific customer type, a particular problem to solve, and a focused product / solution in order to gain maximum signal during your pivot

When Steve Jobs made presentations, he was known to delete all the words on a PowerPoint slide, leaving just the most important one.

If I had to delete every word in this chapter except for two (look I'm not Steve Jobs), here's what it would look like:

Get specific.

The next step of your pivot journey is figuring out what direction to take the company in. Your beautifully, wonderfully dissatisfied users have ranted to you about many problems. Pick one. Then do a really, really good job of solving it.

Staying with Jobs for a minute, the man was obsessed with simplicity and focus. When he came back to Apple in 1997 as interim CEO, Jobs got rid of the clutter. The company catalog went from 350 products to just 10.

By simplifying their product line, they could stop spending time, money, and people on things that weren't working, and where they couldn't win. Instead, they narrowed in with a core lineup focused on just 2 kinds of products (a desktop and laptop), designed for two different audiences (consumers and professionals).

Did it work? Well, Apple, which recorded a billion dollar loss in 1997, turned a cool $300 million profit in 1998[54].

The allure of "everybody"

When you're in the midst of a pivot, it's tempting to get caught up in the grand scale of opportunities. Thinking about how every student, every salesperson, every organization has to use the new shiny thing you're building. And that will hopefully be true one day. For now though, you gotta go ultra-specific.

When my team was pitching ourselves to YC initially, we actually had a different business and it was about nostalgia, basically the memories feature of Photos and Facebook, but YC partners asked who would use it and we literally said "everyone we know", and they were definitely not happy with that.

Don't get me wrong, having big dreams is fantastic. You gotta swing for the fences for your venture-backed startup to make sense. But the first step is drilling down into the nitty gritty details.

---

[54] https://www.vox.com/2014/11/17/18076360/apple

Even Amazon started with one thing

Even the Everything Store (aka Amazon) started out only selling just one product: books. In Bezos' 1997 interview at the SLA Annual Conference, just 3 years after he founded Amazon, Bezos explains why he started with books[55]. He says:

> *"I picked books as the first best product to sell online after making a list of like 20 different products that you might be able to sell. Books were great as the first best because books are incredibly unusual in one respect, that is there are more items in the book category than there are items in any other category by far."*

Bezos then dives into a bunch of stats about active book titles v. music CDs. But here's what's cool — even back then, Bezos (before he hit the gym and embraced the Mr. Clean look) refers to books as the first best product to sell.

It's clear he always wanted to sell more than just books. Bezos had a grand plan to build an all-inclusive online store, but he kicked things off with one thing that made sense. Books were non-perishable, easy to warehouse, and shipped well. The takeaway here is to get laser-focused at the start of your pivot. Amazon went from selling books to being the world's biggest e-commerce marketplace. Get specific at the start, it works.

There are a few objective reasons why locking down a narrow problem for a narrow set of initial users is good:

1. **Easier to build**. If the problem you're solving is small, the less your solution actually needs to do. You don't need a complex solution because a simple fix works just fine — the more straightforward, the better.

---

[55]https://www.businessinsider.nl/1997-jeff-bezos-amazon-empire-viral-video-books-2019-11/

2. **Better results**. With a focused approach to problem-solving, you've got higher chances of hitting the mark. Designing a tailored solution for a precise need is way easier than thinking up an all-inclusive answer for an undefined problem.

3. **More satisfied customers**. Less customers also means less people to make happy. It gives you the bandwidth to realistically invest time and attention to understand their needs. You can build a loyal customer base that'll be more than happy to endorse your product. That makes new- user acquisition a whole lot easier as well.

Narrowing down—nurses and Normandy

The story of tech-enabled healthcare platform Clipboard Health is all about getting specific. Founder Wei Deng's startup journey was driven by her passion to make "the American Dream" accessible to more people.

Her original idea was refinancing student loans for people in exchange for a share of their income once they secured jobs. Wei looked up the largest occupations in the United States (including nurses) and started talking to as many potential customers as she could. In these conversations, she realized two things: one, no one was interested in the refinancing product; and two, nurses, in particular, were actually struggling with a whole different problem.

Nurses had a hard time finding their first job after graduating from school. Wei channeled her broad passion to help people realize their potential into building a specific service to make the lives of nurses more comfortable and secure.

Geoffrey Moore also talks about the importance of starting out with something specific in his classic book *Crossing the Chasm*. He uses the

Allied invasion of Normandy on D-Day as a how-to for building a GTM strategy:

> *"Our immediate goal is to transition from an early market base (England) to a strategic target market segment in the mainstream (beaches at Normandy)...We'll force the competition out of our targeted niche markets (secure the beachhead), then move out to take over the additional market segments (districts of France), on our way to overall market domination (the liberation of Europe)."*

When you're deciding what to pivot to, remember to zoom in on that first, all-important beachhead. As Moore says, "if we don't take Normandy, we don't have to worry about how we're going to take Paris." You don't have a whole lot of time when you pivot, so it helps to stay narrow and gain momentum.

Don't worry if your initial market looks "too small" — you just need to make sure that you're adjacent to a big opportunity. PG reps this POV too[56]:

> *"Indeed, it's often better to start in a small market that will either turn into a big one or from which you can move into a big one. There just has to be some plausible sequence of hops that leads to dominating a big market a few years down the line."*

Building something with potential to branch into banking, CRM, sales, or virtually anything communications-related means that you're on the right track. Get a small part of the big money pie.

Take any gig economy company—Airbnb, Lyft, Uber, Doordash—they all started out by functioning in one specific city before spreading out. Honestly, it's just good practical sense to do that.

---

[56] http://www.paulgraham.com/convince.html?viewfullsite=1

When it makes sense to go wide

Of course, there are times when broadening your scope is a good thing. Sometimes what you've made is waaaay too specific. Or maybe you just haven't figured out who the right customer for your product is. But if you're trying to reach a huge audience, you're going to need a high-visibility event like a PR story or going viral on Product Hunt for it to work out. Eyewear giant Warby Parker supposedly grew off the back of good PR. According to an in-depth case study by consulting agency G&Co.[57]:

> *"One of the three things the DTC brand spent money on in the beginning years was its inventory, its eCommerce storefront, and a public relations firm. And while a PR team may sound trivial to some, the positive press Warby Parker generated due to their early investment was what set it on a trailblazing trajectory."*

So what does all this mean? Simple products are clear, compelling, and differentiated. But there's always a temptation to accommodate more needs, add more features, reach more customer segments. That creates confusion, reduces efficiency of creation/iteration, and increases cost of upkeep. It makes you slow. When you're pivoting, you gotta be quick. Narrow down to stay fast, agile, and responsive.

Once you've set up a small but meaningful target, it's time to hit a bullseye. That might mean getting creative and dare I say a little crazy. In the next chapter we will talk about how to unleash that inner innovator so you can start taking action fast.

---

[57] https://www.g-co.agency/insights/warby-parker-advertising-and-marketing-strategy-case-study

Recap Chapter 10: Narrow Down

The grand opportunity trap

- It's tempting to design a product for everyone
- Don't do it. Even Amazon started with one thing

Cut the clutter—nurses and Normandy

- Jobs's "one word rule" and making $1.3B by getting rid of products
- The story of Clipboard Health - why it's important to get specific
- Author Geoffrey Moore's beachhead strategy in Crossing the Chasm

Don't forget to keep dreaming big

- Start small but stay adjacent to big opportunities - Amazon expanding from books to retail
- Warby Parker's strategy of using high-visibility events for PR

CHAPTER 11

WORK LIKE A (MAD) SCIENTIST

Push the boundaries of your product and marketing to get as much signal as quickly as possible

Lyft was born out of a hackathon.

Founders Logan Green and John Zimmer had opened their enterprise carpooling platform to consumers several months earlier, allowing anyone to offer or accept a ride to a shared destination, even if they weren't affiliated with an academic or corporate partner. But despite their best efforts, they were struggling to acquire users and grow.

So they shot the zombie and stopped building. But now what? At 30 employees, they were larger than a scrappy "a two-pizza" team and this temporary pause could only buy them so much time.

The founders decided to hold an internal hack day, encouraging teams to channel their inner mad scientist and build a variety of prototypes that got them closer to their vision of a collaborative society where transportation was easily accessible and shared by all.

The teams produced 3 ideas:

1. **On My Way**—which showed a driver's ETA to pick up their friend
2. **Journey**—a way to capture and share photos and music from a road trip
3. **Zimride Instant**—the proto-version of Lyft that allowed for real-time ride requests

I met an ex-Zimride employee in Washington D.C who explained the whole hack day to me. During a coffee chat, I asked him about the team's feedback on the different ideas. He smiled and said, "I told Logan I liked the 'On My Way' concept, and he said 'Cool...but I think we're gonna go with Instant.'"

When it comes to inventing the next big thing, you can't be shy, especially in a pivot. This is your chance to make a breakthrough in your product, so don't waste your shot.

Cook up the crazy

Early in the book, I told you about entrepreneur and investor Andrew Lee's advice to us at Ridejoy:

"Ask yourself 'What kind of crazy shit do I want to try before I go under?'"

One of the things we never tried, which I deeply regret, was the "perfect inventory" test. Just let a subset of people experience a fake version of the app with rides that perfectly lined up with your request and see if people booked. It was a crazy idea but would have given us a strong signal if this was worth pursuing.

This chapter was originally called "Think Like a Scientist" because it was about developing hypotheses and testing them. But in the real-world, scientists are pretty cautious, they do a ton of research (lit reviews) on their areas before trying anything, and their goal is to incrementally increase our understanding of the world.

But you don't have time for that. You gotta think (and work!) like a *mad* scientist. We're talking Nikola Tesla, Doc Brown, or Rick Sanchez. What's the difference?

They have the technical chops and the logical mind, but they also aren't afraid of risk and have big ambitions.

Keep swimming

Most entrepreneurs fall into one of two camps when it comes to the prototyping and iteration part of the Pivot Pilot: love it or hate it. Some of you are all about this part of the journey. You come alive and are excited to get crazy and try all kinds of stuff.

For others, you feel rusty and maybe embarrassed about this "step back". Better to see it as retaking a class to get a better grade. I once failed an organic chemistry class after taking it pass/fail. I was told I had to take it for a letter grade for it to have counted anyway and got a C+. Don't be like me—do better!

During a Pivot Pilot, time is of the essence, and the goal is to get as much high quality information as possible. By this point, you've talked to dozens, perhaps even hundreds of current or potential customers in your target markets. You have an understanding of their problems and their worldview.

Now it's time to start doing. Take it from Coinbase founder Brian Armstrong:

> *"When you're in the pre-product market fit stage, the best advice I can offer is that action produces information. Just keep moving forward. Paul Graham had a great line about this, saying that startups are like sharks; if they stop swimming, they die. So even if you're uncertain, take action, because it will provide valuable insights. This has been true in my experience. There were times I decided to act instead of endlessly debating, and although I might have made mistakes, it led me to better ideas."*

Play with the edges

Pushing a new product to an edge on an important dimension is one of the most powerful ways to stand out and capture attention and word of mouth, which you desperately need in the early days of your company. Seth Godin calls this being "remarkable", while others might call it "a unique selling proposition" or a key differentiator.

Tiniest—many products take up too much space (physically or on disk) and being the product with the smallest footprint is a powerful way to get attention. Just look at Preact.js, which offered one of the most popular features (virtual DOM) of one of the most popular javascript libraries (React.js) in a tiny 3kb package and garnered 35k stars on Github as a result.

Fastest (to run)—no one wants to wait and speed is always appreciated by users, especially the speed of slow and/or frequently performed actions. Email service Superhuman doubled down on the speed dimension after hearing from early users (busy execs and investors) who valued every precious second saved.

Fastest (to get started)—many products are annoying because of how quickly it takes to get started. One of the reasons for the Ruby on Rails's massive popularity in the Web 2.0 era (late 2000's to mid 2010's) was how you could build a working blog from a single line (rails new blog) on the terminal.

Largest—sometimes people feel hampered by the lack of space or capacity and if you can pull it off, an order of magnitude increase can win a lot of love. For instance, Google made waves by offering 2 GB of storage free for Gmail when alternatives only gave you 100 MB.

Most open—many companies have succeeded by making their software open source to gain adoption, especially when the alternative is extremely restrictive, like Linux (vs Apple or Windows), Android (vs iOS), and LlaMA (vs OpenAI).

Most customizable—while lots of products try to be easy to use, sometimes what users really want is to let their inner control freak out. Squarespace is an expensive but popular web hosting platform that allows designers to adjust layouts with a drag-and-drop grid system that can be adjusted for desktop vs mobile.

Most compatible—so much of software is integration with other software, and when you have a large array of integrations with other companies, you can capture a lot of business. Segment broke out initially by offering a way to unify 7 different analytics platforms into a single system, and now offers more than 300.

Most specialized—this is what we talked about in the previous chapter (Narrow Down) and one of the easiest ways to push an edge. Sketch became the favorite app for product designers because it prioritized features that helped designers create software experiences (rather than graphic images or print layouts).

Most collaborative—the Internet opened up the possibility for conversation, communication, and collaboration like never before. We talked earlier about the success of Sketch, but Figma then ate their lunch by turning the single-player experience of designing into a multiplayer one.

Most beautiful—people like things that look good. Naturally this makes sense for things that they wear or have in their house, but anything they have to use regularly comes into play. There are so many "new tab" Chrome extensions, but Momentum has over 3M users because they slap a gorgeous new image of a dazzling natural landscape every day.

Define what good looks like

A critical question often arises during pivotal moments in a company's journey: "What's your metric for success? And if you don't hit it, when can we shut this down?"

This question was raised by engineers when the founders of Lyft announced they were pivoting. Such queries get to the heart of a crucial issue—defining what "good enough" looks like. Whether you're launching a product, exploring a new channel, or executing a business strategy, setting clear goals is imperative.

Start your week by identifying a specific target.

For example, you might aim to engage with 10 users and have at least three of them use your product. While securing more users would be ideal, achieving this baseline of three confirms that you're on the right track. If you don't reach this threshold, it's a signal to reevaluate your approach; perhaps the market doesn't want your product, or at least not in its current form. Implement this practice of goal-setting and review rigorously on a weekly basis.

This philosophy of regular assessment was a big part of the program at Techstars Alexa. Every Friday, we'd all gather at a blackboard at the back of the room, crack open some cold ones, and one-by-one each team would review the goals they had set and discuss the metrics they aimed to improve. We'd write these goals down on a blackboard for everyone to see, fostering a culture of transparency and accountability. The discussions that followed served as an inflection point: should the original goal stand, or does a new strategy need to be implemented?

Vanta's many prototypes

Compliance product Vanta founder Christina Cacioppo led the startup through two pivots before finally landing on compliance certification—first a product that put together a best practices checklist designed to make early stage companies security compliant; and then, a B2B assistant which helped companies fill in security compliance questionnaires.

Both times, Cacioppo prioritized getting the product tested by users over everything else. She didn't write a single line of code until she had customers using it and telling her it was solving a real problem. Cacioppo was extreme about this[58], even when it meant waking up at 5:45 AM to manually send out "automated" emails.

In an interview for First Round, she said[59]:

---

[58]https://www.forbes.com/sites/phoebeliu/2023/06/25/how-christina-cacioppo-buil t-startup-vanta-into-a-16-billion-unicorn-to-automate-complicated-security-complia nce-issues/?sh=d2f9e4b16168

[59] https://review.firstround.com/vantas-path-to-product-market-fit

"The easy part is writing code. The hard part is building something people want. So focus first on finding the thing people want."

Limit scope, not quality

When you're developing a prototype or a Minimum Viable Product (MVP), you have two main goals. First, make sure you can technically build your idea. Second, find out if people really want what you're creating. If you're more worried about the technical side, like whether it's feasible to build, then work on that first. But more often, the big question is about demand. Finding out if people want your product is usually trickier than solving how to build it.

As we discussed previously: limit scope, not quality. Instead of trying to make your MVP do a ton of things all at once, focus on making one feature really good. For example, if you're working on an image or video editing suite, pick one thing—like cropping or text captioning—and make it awesome. You want to perfect this one feature, even if it's very specific.

Build credibility through a narrow use-case

Why focus on just one feature? Because if that one thing is great, people will be impressed and trust your brand. This helps build your credibility. If you try to do too much and none of it is good, people won't return to try your next feature. Once you nail the first feature, you can tell your users, "Hey, we added something new!" They may not have needed the first feature, but the next one could be exactly what they're looking for.

By doing this, you create a strong MVP that has a better chance of impressing people and succeeding in the market.

You need to know if you can reach the right people for your product. See if you can find the best way to reach the right people (asking for referrals, engaging in forums / communities, cold outreach, newsletter ads, etc).

Learn from your failures

When you're in the experimental phase of a startup pivot, you're constantly testing new prototypes and ideas. Silicon Valley often says failure is good, or that second-time founders are better because they've learned from mistakes. But is there proof?

Dashun Wang, a business professor, tackled this question in a paper published in Nature[60]. He and his team used math models to study three data sets: NIH grants, startups with venture funding, and terrorist attacks. They looked at the first and second-to-last attempts of each group to understand if they improved over time.

One major finding was that *luck isn't the main factor in success.* Instead, people who were successful in the end did better in their later attempts than their first. This means they were learning or getting better at what they were doing. It might not all be learning; maybe they got more resources or funding, but improvement was noticeable.

The study introduced a useful concept for startups: *keep what works.*

Every innovation attempt has multiple components. Even if the whole prototype failed, some parts might be good. Successful teams reused these good parts in their next attempt, showing they've learned from previous failures.

---

[60] https://www.nature.com/articles/s41586-019-0941-9

Wang actually found that he could predict someone's learning ability by measuring the time between their first few attempts. If they tried again quickly, they likely reused some components, showing they learned something.

The takeaway is clear:

- It's not about avoiding failure, but about learning from every prototype and test.
- Whether you're seeking venture funding or simply testing a new feature, don't scrap the entire project after a miss.
- Take the good parts, learn from them, and use them in your next attempt.
- This iterative process is the real pathway to long-lasting success.

How Houseparty iterated their way into teenagers's hearts

Viral pandemic app Houseparty started out as a way for people to broadcast live video to their Twitter feed. The broadcasting play got instant traction, but numbers started falling when Facebook and Twitter started building live streaming tools of their own.

The startup came up with the idea for Houseparty—an app that started broadcasting using the device's front camera as soon as it was opened, while also notifying friends that you were live—while on a company retreat. Since it was targeting a younger demographic, the COO visited many schools to get real user feedback on the app. Their engineering goal was to tweak the app and have a new version ready at the end of every week. It was through this (and schools closing to reduce transmission) that the company reached 50 million sign ups in a single month[61].

---

[61] https://review.firstround.com/vantas-path-to-product-market-fit

Building the Impossible with ClearBrain

After a PM stint at Optimizely, Bilal Mahmood started ClearBrain focused on data integration, helping analysts reduce the time to ingest and process data from various sources. But after his technical co-founder left, Bilal had to find a new path forward for the business.

While ClearBrain had signed letters of intent from customers, they struggled to convert those into contracts or active users. Bilal realized that customers were less concerned about data integration itself and more focused on the end result of leveraging that data to train ML models.

So Bilal found a new technical co-founder and they began offering custom ML models, analyzing each customer's existing data warehouse. This led to more letters of intent and eventually an Airbnb contract worth $500k.

But when growth stalled at $1M ARR, Bilal again saw limitations to this customized approach. Customers kept asking how to differentiate correlation and causation in their data. Though Bilal thought it impossible at first, his engineers convinced him that they could build a predictive analytics platform that would bypass the need to run extensive A/B tests.

After 6 months developing this new capability, Bilal was amazed that his team had made the impossible possible. They had built the first large-scale causal inference engine to allow growth teams to measure the causal effect of every action[62]. This finally resonated with customers, driving a surge in usage and interest and ClearBrain was eventually acquired by Amplitude in 2020, just before the pandemic.

---

[62] https://blog.clearbrain.com/posts/introducing-causal-analytics

Though the path was winding, Bilal learned to trust his team's unconventional ideas over his own doubts. This openness paved the way for his team to unlock an innovative solution that solved his customers' biggest need.

Building fast and learning faster is stressful. In fact, everything about doing a pivot, everything about doing a startup can be demanding. To continue executing at a high level while making great decisions, you're going to need to cultivate resilience. In this final chapter, I'll share with you the strategies that can keep your head on straight and burnout at bay.

Let's go.

Chapter 11 Recap: Work Like a (Mad) Scientist

Think about the crazy shit you want to try before going under

- Lyft's origin story was a company hackathon
- Don't just think like a scientist. Think like a MAD one
- And start doing!

Play with the edge cases, then pick one

- Godin calls it being "remarkable," other people call it a USP. This is about how you can find it or be it
- Tiniest (Preact.js), Fastest to run (Superhuman), Fastest to get started (Ruby on Rails)
- Most open (LlaMA v. OpenAI), Most customizable (Squarespace), Most compatible (Segment), Most collaborative (Figma)

Define what "good" looks like to measure your progress

- The blackboard at Techstars - setting weekly metric goals
- Vanta case study from B2B Alex to compliance certification. Cacioppo woke up at 5:45 AM to manually send "automated" emails
- Tips for a killer MVP. Limit scope not quality.
- Data-backed research says you should iterate fast. That's also how Houseparty went viral
- ClearBrain (acquired by Amplitude): data integration to custom ML models

CHAPTER 12

STAY RESILIENT

It's not enough to survive the pivot, you must come out stronger—so prioritize your physical, mental, and emotional well-being

No one can better attest to the enormous strain of a pivot than Lyft cofounder John Zimmer—whose pivot-related migraines landed him in the hospital.

During the pivot from Zimride to Lyft, there was a lot of uncertainty about how the various products and business lines would be handled. The stress of not making decisions led to Zimmer experiencing terrible migraines, stabbing pain behind his eyes that prevented him from working or doing much of anything. He sought help from his primary care physician and even went in for a brain scan[63] to determine the root cause of the problem. But despite all his efforts,

---

[63] https://techcrunch.com/2014/08/29/6000-words-about-a-pink-mustache/

the splitting headaches got worse and no one could give him a good reason why.

It was ultimately the decision to make Lyft the future of the company and deprioritize Zimride that finally lifted Zimmer's month-long torment. This shit ain't easy and don't let anyone tell you otherwise. We'll hear more about Zimmer's journey with mental health and why Zimmer has become an advocate for investing in one's physical and mental health, and we'll learn more about his story in this chapter.

To be a founder is to put yourself through immense hardship and struggle in service of a greater mission or vision. Building a company takes a toll on you physically, mentally, and emotionally as you try to hold everything together and make progress with limited resources and immense uncertainty. This is doubly true during a pivot.

This chapter is all about how to take care of yourself during a pivot.

Understand that a pivot can be an overwhelming experience. The familiar markers of success are gone and product momentum is minimal. You're no longer pushing hard towards the next milestone. Instead, you're plunged into an unfamiliar realm of heightened uncertainty. This isn't your usual startup stress of having tons to do or adapting to rapid changes—it's the unsettling feeling of standing at a crossroad, unsure which path to take.

UT Austin psych professor and executive coach Gena Gorlin writes about the idea that founders and builders are actually an underserved population given the challenges of what they are trying to achieve[64].

---

[64] https://builders.genagorlin.com/p/raising-humanitys-psychological-ceiling

"The more ambitious and innovative your life projects, in sum, the more formidable your psychological needs—and the fewer the psychological resources that have been developed for navigating those needs. But this doesn't mean you either have to settle for misery and burnout or lower your ambitions. Rather, it means you need to be that much more vocal in articulating and advocating for your needs, and that much more entrepreneurial about hunting down the best available resources and bootstrapping them to suit your specific psychological purposes."

For the sake of your company, your customers, your family, and yourself, it is critical you prioritize your own well-being. You're the captain of the ship, no one can do it for you.

That said, there's still a lot of areas where you can make investments.

Pay down your body's tech debt

I find that many entrepreneurs, especially technical ones who value the power of their mind, tend to be incredibly ignorant of their bodies. You are not a brain housed inside a body that inconveniently has to eat and go to the bathroom. You are primarily a physical creature, a complex set of interconnected biological systems that happens to be able to think abstract thoughts.

Addressing your physical needs and nourishing your physical body can transform your ability to build and lead. Let's start with the most challenging one for busy founders to accept: sleep.

Sleep: Snoozing like a champion

While late meetings or coding sessions are sometimes inevitable, do your best to prioritize high quality sleep. Sleep is one of the most important ways our brains recover from the flood of information and activity we engage in all day long. Poor sleep affects your brain's ability to consolidate memories and make new connections.

You've heard it before, but try to get seven or more hours of sleep a night. Seriously. The last hour of work tends to be your least productive one, and better rest will make the hours you do work far more effective.

Sleep experts tell us that it's not just about sleep duration, but quality and timing. Investing in blackout curtains or a comfortable sleep mask with raised eye cups makes the sleep you do get more restorative. If you live in a noisy area, pick up a white noise machine or find a playlist on your phone and set up a sleep timer. They also now make reusable ear plugs that stay in your ear even when you can lay on your side that can help seal off the rest of the world.

Studies show that sleeping at the same time every night is ideal. Many entrepreneurs, myself included, find routines hard but make this one a priority. Create a winddown routine that includes silencing your phone, turning off screens, and stretching, reading, or meditating to relax your mind. This will help train your brain to ease out of work mode and into rest mode.

"But Jason, I don't need this stuff," you say. "I'm built different." Maybe you're part of the 5% who can sleep 4 hours a night and be fine, but prove it to yourself with data. Get yourself a fitness/sleep tracker that you can wear to bed. Track how your sleep duration, timing, and things like alcohol, caffeine, and screen time affect your sleep quality and sense of restfulness in the morning. Odds are, poor sleep affects you more than you think.

Sleep has been studied extensively over the last few decades and has overwhelmingly shown powerful effects on people's cognitive, emotional, and physical performance. Pro teams in the NFL, MLB, NHL, and NBA work with sleep experts to help their players get that extra edge. According to Cheri Mah, UCSF sleep researcher and sleep consultant to the Golden State Warriors, sleep can make even the world's most elite athletes that much better[65].

> *"The comparison most of us make, when talking about the importance of sleep, is to performance-enhancing drugs. But when you look at the research, it suggests a solid foundation of rest and recovery is worth way more than [a performance-enhancing drug]."*

Exercise & Outdoors: The power of touching grass

Our bodies were meant to move, not sit at a desk or couch for 8 hours a day. It's truly one of the greatest cheat codes on Earth because when you exercise to a moderate degree of exertion (aka sweating), your body releases chemicals to make your brain happier and smarter.

Exercise gives you happy juice

You've probably heard of endorphins, the class of neurotransmitter released during exercise, but did you know what they stand for? ENDO-genous mo-RPHINE. This stuff is literally your own personal stash of drugs that get released during a workout, as well as eating, massage, and sex, making you feel less pain, less stress, and improve overall well-being.

Sweat makes you smarter

But there's more. Exercise also releases brain-derived neurotrophic factor, a protein that aids in neuron survival and growth, which helps you learn faster and make new connections. This release doesn't just make you sharper in the moment—immediately after the exercise, but compounds over time: one study of Canadian adults found significant cognitive improvements in adults after 4 months of aerobic and circuit training (150 mins a week)[66].

---

[65]https://www.wired.com/story/how-science-helps-the-warriors-sleep-their-way-to-success/

[66] https://www.healthline.com/health-news/exercise-can-make-you-smarter-102912

Make it your own

The key is to find something that works for you. If you like running, then run. If you hate running, do something else!

Try Starting Strength, Apple Fitness, CrossFit (with a good instructor), spin classes. Working out VR workouts can actually be quite fun and there is literally an endless supply of free guided workouts on YouTube (try "20 mins beginner bodyweight workout"). Walking meetings can be a great way to get some movement while catching up with a colleague or close partner.

As a former NCAA athlete and ADHD entrepreneur, I've consistently prioritized exercise and movement because it's the only way I can feel settled enough to do cognitively demanding tasks. But don't take my word for it, here's Stanford instructor Kelly McGonigal in her book *The Joy of Movement*:

> *"People who are regularly active have a stronger sense of purpose, and they experience more gratitude, love, and hope. They feel more connected to their communities, and are less likely to suffer from loneliness or become depressed. These benefits are seen throughout the lifespan."*

The power of the outside:

Let's pivot to a setting that's as essential as your movement: the great outdoors. We don't call it "touching grass" for nothing. Studies have found that spending time in nature, even a small amount, can reduce stress hormone levels, lower anxiety, improve mood, and boost concentration.

You're not just admiring the greenery; you're bathing in an environment that's been scientifically proven to improve your mental well-being. It's a concept the Japanese have termed "Shinrin-yoku," or "forest bathing," which entails immersing oneself in nature and soaking in the environment through your senses. This isn't some woo-woo concept either; a study published in Environmental Health and Preventative Medicine found that forest environments promote lower concentrations of cortisol, lower pulse rate, lower blood pressure, and greater parasympathetic nerve activity than do city environments.

The sunshine factor

Sunlight, the star of the outdoors, isn't just there to give you a tan. Exposing your skin to sunlight allows your body to produce Vitamin D—a critical nutrient for mental and physical health. Vitamin D deficiency has been linked to mood disorders and cognitive impairments. You can buy a sunlight lamp from many online retailers if you live in an always cloudy place like San Francisco or London. And of course, use sunscreen if you're going to be outside for a long time. But when managed well, sunlight is a powerful mood-booster that costs nothing.

Get creative

Here's where the real magic happens: Combine exercise with being outdoors. Whether you're going for a run in a nearby trail or playing fetch with your dog in park, merging these two powerful elements amplifies their benefits.

Also, consider walking meetings, not just as a different way to catch up with colleagues, but as a mental and physical revitalization strategy. Most people are fine with going video off and making it an old fashioned phone call, and this works even better in person.

Your body and mind are your most valuable assets. Just like you wouldn't skimp on investing in a robust infrastructure for your startup, don't cut corners on investing in yourself. Engaging in regular physical activity, especially in the nurturing embrace of nature, isn't a luxury—it's a necessity for peak performance and long-term resilience.

Invest in creative hobbies

Beyond just physical health, it's critical to invest time and energy into creative hobbies during a pivot. Creative pursuits like writing, making music, dance, painting, or sketching give your brain a break from the constant work grind.

Ideally, choose hobbies that engage you in different mediums and environments from your day job. If you stare at code all day, get a sketchpad and draw with pens and pencils instead.

Creative hobbies relax the mind and activate what neuroscientists call the "default mode network." This allows the brain to make novel connections and stumble upon creative insights. When you give your brain a break from focused problem-solving mode, it can synthesize information and generate new ideas in powerful ways.

Many of my founder clients use artistic hobbies to explore interests outside work—several made music using DJ mix kits or digital sound stations. This created a sense of freedom and space so they didn't feel trapped and overwhelmed by their company struggles.

When you get perspective, it's easier to stay motivated to convince your team, investors, and customers to stick with you through uncertain times.

Whether it's making art, playing in a recreational sports league, or building stuff in your garage, find whatever creative outlets resonate with you. Use them to remember that you are more than just your work. These practices help cultivate joy and a sense of inspiration that carries you through difficult periods.

Connect with a support network

For years, Greatist and Ness founder Derek Flanzraich organized a casual monthly series at his apartment in New York City called "Scotchpreneur". The idea was simple, bring entrepreneurs together to share a fancy bottle of Scotch, eat snacks and answer a simple question: **What's the your biggest challenge?**

Easy problems weren't allowed. People talked about running out of money, having to fire a problematic employee, deciding to pivot, and occasionally personal challenges like a parent with dementia or a relationship ending. It was an incredible lifeline for me and many of the regular attendees who listened, asked questions, called each other on our bullshit, and offered advice and support during and after the session. Sadly the pandemic ended our in-person sessions and Zoom didn't quite do it. Many members left NYC during those years and the group never reformed.

Seek out your own group of peers and fellow founders where you can be truly vulnerable and talk about difficult topics. Many Y Combinator alumni make a point to meet regularly after graduating the program to share struggles and support one another.

It's also important to build professional support like a therapist, coach, or mentor. As a founder, I worked with an executive coach to help me lead effectively while coping with stress.

Each plays a unique role:

1. **Consultants** give tactical advice in specific domains like marketing strategy, pricing, design, etc. You engage them for specialized project help and pay by the hour or retainer. Use consultants sparingly to solve pressing problems.

2. **Advisors & Mentors** offer high-level strategic guidance based on years of experience building startups. Investors often serve this advisory role. Schedule check-ins every month or quarter to get their wisdom on major decisions.

3. **Coaches** help you manage leadership challenges and personal growth. They create a judgment-free space to process the mental toll of leading a startup amidst uncertainty and give

you context as they have seen how other clients deal with these challenges. Set up a regular cadence of sessions to unload stress and gain context and self-awareness.

4. **Therapists** provide mental health support and tools to overcome unproductive thoughts. But they may not understand the realities of entrepreneurial obsession the way a coach does. Seek therapy if you feel depressed, anxious, burnt out or have other mental health concerns.

The right support system allows you to be vulnerable, speak candidly about challenges, and get encouragement to stay the course. It also exposes you to patterns across many companies, so you don't feel alone in the struggle.

Recall what Dr. Gorlin said at the start of the chapter, that the psychological needs of founders are far greater than the general population, meaning seeking professional, far from a sign of weakness, is simply a smart problem-solving tactic, just as hiring a specialist to deal with a part of your business might be.

John Zimmer learned this the hard way during a bout of depression he experienced several years into the Lyft vs. Uber battle. The company had less than 5 months of runway and the pressure took him to a dark place[67].

> "It could manifest in having trouble getting out of bed in the morning. It manifested in me burying myself in solving the work so that I could unblock the depression. You get stuck in this dark cloud, and this exhaustion and this cycling of bad thoughts: 'Will this survive? Am I not good enough?'"

Zimmer had close friends who got him into therapy, which helped destigmatize his depression. With a combination of therapy,

---

[67]https://people.com/human-interest/lyft-co-founder-john-zimmer-reveals-mental-health-struggles-talks-work-to-destigmatize/

medication for a time, and hard sweaty bouts of Crossfit and boxing workouts, he was able to get back on his feet.

Leaning on others for wisdom and comfort renews your spirit during an exhausting pivot. You don't need to shoulder the entire burden alone.

Friends, family, and romantic partners

While entrepreneurs tend to gravitate toward professional support systems, don't neglect your family, close friendships, and romantic relationships during a pivot.

Personal relationships can provide validation and comfort when your worth feels tied to your struggling startup. They remind you of your inherent value as a human being.

That said, your nearest and dearest may not fully understand the realities of your situation. The obsessive, rollercoaster lifestyle of entrepreneurs is foreign to most.

But even if they can't grasp the details, they can still be a sympathetic ear. Venting your frustrations can be cathartic.

Lean on those closest to you, especially relationships outside your field and social circle. The outside perspective helps ground you. Don't let work chaos distance you from people who knew you before you were a founder.

Make time for quality bonding away from work talk. This space allows you to engage different parts of yourself and prevents burnout. You'll return to your startup refreshed.

Cut yourself a tiny bit of slack

Don't be too hard on yourself throughout the process—it's not warranted and it's ultimately unproductive.

By deciding to pivot, you've shown great courage and conviction to challenge the status quo. You recognized the need for change and were willing to go against biases like sunk cost fallacy and the endowment effect. That alone deserves praise.

Whether or not this pivot succeeds, know that you made the best decision you could with limited information. Hindsight is 20/20 - judge yourself based on acting on your principles, not the outcome.

The startup world forgives failure when it comes from bold ambition. Look at the most successful Silicon Valley entrepreneurs and VC's - they have failures scattered amongst their wins. And many companies that started out strong eventually fade—our community understands that breakouts success is rare, dominance is fleeting, and failure is commonplace.

Success breeds failure, and vice versa. Luck plays a major role.

Control what you can - making sound decisions based on your current knowledge. You may get this pivot right or wrong, but either way, you'll have more chances down the road.

Remind yourself that you're getting another shot with this new direction. Even if this startup fails, you will take the lessons learned here to your next venture. Your learnings compound over time.

So cut yourself some slack during the difficult process of pivoting. Persist, adapt, and know that better times lie ahead.

Recap Chapter 12: Stay Resilient

Don't just survive the pivot. Come out stronger

- Lyft cofounder John Zimmer on the mental and physical toll of pivoting
- Founders have different psychological needs. Learn about them. Take care of yourself

Pay down your body's tech debt

- Get your sleep. That's all
- Make sure you're touching grass - be active and get outside

Find your creative outlet

- Invest in old hobbies that have nothing to do with work. Paint, sketch, dance
- Keep looking for new ones

Cultivate a strong support network. Show up for yourself and others

- Organize a "Scotchpreneur" style meetups
- Seek out your loved ones and lean on them
- Cut yourself a tiny bit of slack

APPENDIX I

VENTURE MATH

Why nothing less than a billion dollar outcome is going to cut for your investors

As much as we might envy our venture capital partners—sitting on what seems like massive amounts of capital while making decisions about who to fund, the truth is far more complicated. It's tough to succeed as a venture capitalist, especially now that we've passed the 2020-2022 funding bubble.

Why Your Investors Need a Billion Plus Outcome

For seed and Series A VC funds, having a portfolio company deliver a $100 million exit is not enough. They're looking for at least one $1+ billion return to make their own business work.

Why do VC's care so much about getting such a large return from their portfolio companies? Here's the math to back it up:

When you raise money from a VC, they're aiming to give their limited partners (LPs) a 3x return. The logic behind the 3x? VC funds tend to run for about 10 years. On an average, stock markets give returns of 9-12% per year—and the VC should match or beat those returns for it to make sense for the LP to invest. 12% per year over 10 years is ~310%.

This is ignoring the 2% fee + 20% carry that most funds take, which further increases the return the VC must deliver in order to make their investment vehicle attractive to LPs.

Let's take a VC with a $100 million fund who is investing $10 million in 10 companies. The VC is gunning for exits that will make them at least $300 million.

Now, if half of the 10 startups sell for $100 million, the VC would make ~$25 million each. With only $125 million in the bank, they haven't even delivered half of the expected 3x return.

On the other hand, if even 1 of those companies hits the $1 billion mark, the VC pockets $250 million. They just need one or two more $100 million companies and they can call it a day.

And of course with the rise of decacorns ($10+ billion companies), investors are continuing hunting for ever larger and rare possibilities from their portfolios.

According to Chris Dixon, another a16z partner, 6% of investments representing 4.5% of VC dollars invested generated ~60% of the total returns[68]. Firms that don't have at least one billion dollar exit will likely fail to satisfy the expectations of their limited partners (LPs).

---

[68] https://a16z.com/performance-data-and-the-babe-ruth-effect-in-venture-capital/

Appendix II

PIVOT CASE STUDIES

Every pivot is a little different, and many pivot stories lack context—here's true story behind 5 iconic pivots

GOAT (Grubwithus)

Shared dinners → Sneakerhead heaven

Original premise
What problem were they solving? For who? What was the intended solution?

Eddy Lu and Daishin Sugano had just moved to Chicago when they realized that after college it was difficult to make friends in a new city. Recognising this as a common problem for adults, they built Grubwithus to introduce users to new people over group meals in local restaurants. A user could either create their own or join an already scheduled meet-up at one of the startup's partner restaurants by paying a fixed price upfront. This price included a $2 to $3 service fee per reservation which Grubwithus pocketed.

Align

Warning Signs

What made them think they needed to pivot from their original premise?

They realized that it was difficult to scale this business model because it was easy for users to cancel on strangers at the last minute—people often had to work late, had conflicting plans, or just ended up being too nervous. On the other hand, coordinating with restaurants was becoming complicated from an operations perspective. Lu said that a major red flag for him was that he didn't enjoy working on the project anymore.

Time to Pivot

How long after they started the company did they think about pivoting?

They started thinking about pivoting from Grubwithus around 4 years after they started the company. They briefly relaunched the platform as "Superb", an app to make lists of places, before finally starting GOAT in the summer of 2015.

Initial Funding / Team

Had they raised money before their pivot? How much? How big was the team?

Yes. Lu and Sugano had raised a total of $7.7 million for Grubwithus and had a team of around 12-15 employees.

Explore

New Opportunity

What were the key insights that drove their pivot?

They stumbled upon the business model for GOAT when Sugano had a bad experience ordering a pair of Air Jordan sneakers from Ebay—the kicks he ordered ended up being fakes and he couldn't get his money back. After brainstorming with Greg Bettinelli, their investor from Upfront Ventures, the duo saw that not only was the growing sneakerhead market responsive and young, but the industry was plagued by two problems: authenticity and organization. Lu and Sugano knew that they could leverage technology to solve most of these problems.

Pivot Validation
How did they validate those insights?

In the first few months after the launch, GOAT did around $14,000-20,000 in GMV per month. Sugano would use some of the cash generated to buy sneakers to seed the platform. They hit encouraging milestones like the first time they did $1,000 in sales in a single day, or the day they sold a record high of 5 pairs of sneakers fairly quickly. On their first Black Friday, GOAT offered premium sneakers at very dramatic mark-ups. This went viral. Even though their site crashed almost every day in the week leading up to the sale and they didn't have enough inventory to meet all the orders placed, it put them on the map.

Commit

Length of Pivot
How long did it take to get a clear signal the pivot was working?

In a little over a year since it was launched, GOAT became the leading player in the sneaker marketplace—20,000 sneakers were sold every month.

Pivot Funding
Did they raise money during or after their pivot?

In August 2016, around a year after pivoting, GOAT raised $5 million of venture money in a round classified as Series A-7. The investment was led by Matrix Partners, with participation from existing investors Upfront Ventures and Webb Investment Network. This brought their total capital to around $12.6 million.

Pivot Outcomes

What happened after the pivot? Where is the company today?

GOAT got to PMF in the sneaker marketplace pretty fast. As market leaders, they distinguished themselves by focusing on authenticity. They made sure the fakes were weeded out. Once they had the basic model nailed down, they started broadening their horizons. In 2018, they expanded into brick and mortar with the acquisition of legendary sneaker retailer Fight Club.

A year later their offerings diversified to include apparel and accessories through direct partnerships with brands like Balenciaga and Versace. GOAT has long since grown out of its identity as a reseller. It is positioned as a platform curating top trends in fashion, style, and culture. They also publish a 150-page biannual magazine called GREATEST to better express their views. On the tech side, they've played around with AR try-on features to give users a richer experience. The growing company also raised $196 million in Series F money at a valuation of $3.7 billion in July 2021.

Lessons Learned

1. **Be open to starting over.** Lu and Sugano founded a string of different businesses (including selling golf apparel and buying a creampuff franchise) before GOAT. Almost all of them failed. But the pair's attitude while starting each one was the same—if we believe in the idea, we're going to give it a shot. They knew they could just try the next idea if things didn't work out. Giving up altogether was not an option. And eventually this never say die mindset paid off. As a founder, the chances of getting everything right the first time round are low. Resilience is key here—you have to be willing to go back to the drawing board.

2. **Don't make the same mistake twice.** With Grubwithus, they launched across different cities early on. This quickly backfired because they ended up having to solve the same problems over and over again. Leveraging their past experiences while launching GOAT, Lu and Sugano decided to start small and double down on quality. Your business may have failed, but what you've learned is invaluable. Make sure to use it the next time you give it a go.

3. **Know the details before you decide.** Lu and Sugano extensively researched the sneaker marketplace before starting GOAT. They wanted to know the real pain points, existing solutions, and gaps in the industry. They went down reddit rabbit holes, stalked blogs on the topic, and followed the right social accounts. A deep understanding of the industry is crucial. Do the research before diving in.

Lyft (Zimride)

College carpooling → Real time ride-hailing

Original premise

Logan Green and John Zimmer wanted to reinvent the way in which people got from one place to another. Stuck in traffic jams surrounded by cars with just one person sitting inside, they saw the empty seats as unused inventory which they could leverage. On the heels of Facebooks' success with college students, Zimride initially offered long-distance ridesharing to campuses. The company went on to build a business as a carpooling platform for universities and companies.

Align

Warning Signs

What made them think they needed to pivot from their original premise?

Zimride had built a reliable revenue stream from its enterprise clients (corporates and colleges). But since the founders' had a vision to increase occupancy rates in cars, they decided to make the carpooling service available to individual consumers. Even after months of trying different ways to drum up regular users, they didn't see the growth they were hoping for. Having just launched a mobile website, the next planned course of action was an app. But the engineers were skeptical about building it because everything they'd made so far hadn't amounted to much, in terms of dollars or repeat users. That's when Green and Zimmer decided to stop building.

Time to Pivot

They started building Lyft 5 years after Zimride.

Initial Funding / Team

Zimride had raised a total of around $7.5 million dollars from investors like Mayfield Fund, Floodgate, Keith Rabois, K9 Ventures, and fbFund. They had a team of about 30 employees.

Explore

New Opportunity

In the uncertainty that followed the decision to stop building on Zimride's consumer segment, the idea for real time ride-sharing was born out of an internal hackathon. The original idea was called "Zimride Instant" which connected users calling for rides on their mobiles with members of the community who were available to drive them around. Something like Uber's model without using licensed drivers.

Interestingly, it was the founders' original premise for the company — to increase occupancy rates in cars — that pushed them to pivot. They decided to house it under a separate brand, "Lyft". Another core objective was to market each ride as a social experience — having the passenger sit upfront and starting each ride with a fist bump.

Pivot Validation

Lyft's launch didn't go as planned because a publication broke the press embargos which are infamous amongst Silicon Valley journalists. TechCrunch actually wrote about how the launch was ruined, before eventually publishing an actual piece on it a few days later. But there really is nothing like bad publicity. The demand for Lyft was soon through the roof—in a couple of months they had to add a waitlist to keep up. Their performance metrics like repeat rates and frequency of use were undeniably positive.

Commit

Length of Pivot

Just a few months after it launched, Lyft was generating enough business to be the primary focus of the company. They eventually re-incorporated it into a new company, Lyft Inc., and sold Zimride to Enterprise Holdings.

Pivot Funding

Lyft raised a total of $75 million in a Series B and C one year after pivoting. In the second year post-pivot, the company raised $250 million in a Series D round backed by new investors like Alibaba and Third Point, with participation from existing ones including a16z and Mayfield.

Pivot Outcomes

Lyft went public in 2019 at a valuation of $24 billion. Its value dropped to around $13 billion in December 2021 and then further down to $7 billion as of August 2022 with the rest of the market. Lyft's numbers in Q2 of 2022 have been encouraging—its active ridership and revenue-per-rider numbers have increased. They are experimenting with new features to attract more drivers. Their prudent internal cost cutting measures (slashing travel and expense budgets, cutting back on hiring, shutting down their in-house car business, etc) to combat inflationary pressures also mark the return of operations as a somewhat lean startup.

Lessons Learned

1. **Stand up for your ideas.** With their ridesharing business for enterprise clients doing well, when Green and Zimmer wanted to try carpooling services for individual consumers, a trusted mentor told them it wasn't a good idea. Even though it was difficult to go against his advice, they ended up

pivoting anyway because they believed in it so strongly. They went with their gut because they had real knowledge of the space and saw a huge opportunity. If you're passionate about an idea and you've done the research or have the experience to back it up—stick to your guns even if the people you respect tell you that it's not worth the risk.

2. **Get comfortable with things changing.** The idea for Lyft came from an internal hackathon when Green and Zimmer weren't sure which direction to take the company in. According to Green, building a culture which inspired experimentation was crucial to Lyft's success. While developing these team dynamics, it's also important to be open to pivoting as a founder—chances are that the company you start with will change dramatically, and that's okay.

Nuvocargo (The Lobby)

Coaching job candidates → Cross border shipping

Original premise

Working as an investment banker at Merrill Lynch, Deepak Chhugani believed he was truly lucky to have broken into Wall Street without an Ivy Leage education. He noticed that competent people with similar backgrounds weren't able to land the coveted finance jobs because they didn't have access to the right networks. After quitting his job, he built a startup called The Lobby which connected job seekers with Wall Street 'insiders' for personalized advice and coaching over a phone call.

Align

Warning Signs

A few months after raising $1.2 million for the company, Chhugani realized that its revenue was nowhere near as high as it should have been. He was having a hard time scaling the business model. In the fall of 2018, he gave himself an ultimatum: if the company didn't hit $60,000 in monthly revenue by the end of the year, he would cut his losses and shut the company down.

Time to Pivot

The Lobby was founded in May 2017. Chhugani started thinking about pivoting toward the end of 2018.

Initial Funding / Team

Chhugani had raised $1.2 million pre-pivot and had a team of 8 employees.

Explore

New Opportunity

Chhugani's passion to bridge the gap between the U.S. and Latin America was one of the driving factors behind the pivot. In Ecuador, his father had worked in B2B logistics, connecting Latin American buyers with Asian sellers. While building Nuvocargo, he wanted to leverage his unique advantage of having relationships in Latin America while also understanding western markets. He also saw that it was a traditional industry ripe for disruption.

Pivot Validation

Even with little marketing and a basic sales strategy, Nuvocargo gained traction very quickly. In just a month after its launch in August 2019, the company had already exceeded the highest monthly revenue ever earned by The Lobby. Freight forwarding has many different elements like customs brokerage, insurance protection, and trade financing—by offering all of them in a single platform, Nuvocargo built strong network effects.

Commit

Length of Pivot

18 months after Nuvocargo was launched, Chhugani wrote that the company was at an "inflection point." Nuvocargo consistently showed a > 30% MoM growth in revenue since it started. It also 25xed its revenues in 2020. They had a team of 35 employees and were expecting to grow to a 100+ headcount in the coming year.

Pivot Funding

Nuvocargo has raised a total of $37.3 million since it was launched. In its last round in December 2021, it raised $20.5 million led by Tiger Global.

Pivot Outcomes

Nuvocargo has been growing exponentially in the U.S. and Mexico with a 5x revenue increase from 2020 to 2021, and the number of monthly shipments through the platform increasing by almost 3x in the first six months of 2022 as compared to the same time period in 2021. The company definitely seems to have gotten to PMF. Recognized in YC Top Companies 2022, it has also been ranked the second most innovative company in Latin America by Fast Company. Nuvocargo has also been focusing on launching products and services to streamline invoicing and payment of carriers. An example of this is QuickPay, a cash advance product for carriers in the Nuvocargo marketplace.

Lessons Learned

1. **If it isn't working, shut it down. And then try again.** Chhugani had the foresight to recognize that The Lobby was not working very quickly. When he saw that the monthly revenue numbers didn't meet the deadline he had set for himself, with VC money still in the bank, he decided to cut his losses and try again. As a founder, it is difficult to start over when you've worked hard, raised funds, and hired a team. But it's absolutely crucial not to burn money, time or energy on a bad idea. When executed right, great ideas work quickly—set clear targets for the company, and if it isn't working, have the courage to scrap the idea, take the lessons you've learned, and start again.

2. **We all have an unfair advantage in something. Use yours.** When he built The Lobby, Chhugani was trying to build a company to solve a problem that he had faced himself. "Solve your own problems" was the popular advice doing the rounds at the time. Chhugani realized that as an

urban 20-something millennial who had worked at a bank for 2 years, he would end up trying to solve the same problems as everyone else out there. After looking into the startups that were doing well, he realized it made more sense to choose an area in which he had a tangible edge. When you're picking a problem to solve, make sure to leverage your own unique advantages, find a big market, and have a sound business model.

3. **Nothing can substitute real world work experience.** Chhugani's general advice to anyone trying to break into the startup world would be to start working at another startup. According to him, this is the fastest way to skip the early challenges of being a founder (like building relationships with VCs or other founders, having a good network of lawyers, designers, engineers, etc.).

Houseparty (Meerkat)

Livestreaming → Video chat

Original premise

Ben Rubin and Itai Danino were working on a startup called Yevvo which was originally developed as a tool to let users share live videos with their friends. After a year and a half, it had amassed many random features (like location tagging, text comments, and multiple login options) and a set of largely inactive users. One of the company's investors advised them to focus on doing one thing well. Taking his advice, they shut down Yevvo and eventually launched Meerkat in March 2015—a simple app that allowed users to broadcast live video to their Twitter followers. When the broadcast started, a tweet notified all their followers who could then watch and comment on the live video.

Align

Warning Signs

Rubin began seeing worrying usage statistics on Meerkat—the frequency at which normal users were broadcasting through the app was reducing. In other words, the gap between each time a user launched a broadcast was increasing. Adding to the company's problems, Twitter—which had recently acquired a competing live streaming app (Periscope)—blocked Meerkat's access to its social graph. Facebook also built a real time broadcasting feature into its mobile app. For Meerkat this meant that the celebrities and influencers which formed a bulk of its frequent users would migrate to Twitter and Facebook. Rubin realized two things: Firstly, while broadcasting features were a great addition to already existing networks, it could not justify a whole new medium. Secondly, users couldn't easily engage with each other inside the app.

Time to Pivot

Rubin started thinking about ways in which the startup could pivot around six months after Meerkat was launched.

Initial Funding / Team

The company had raised around $14 million from investors including Greylock and Josh Elman. The team was around 20 employees strong.

Explore

New Opportunity

While on retreat, Rubin asked his team which part of Meerkat they liked using. Most of them said that they enjoyed live streaming when their close friends or family participated. Building on this feedback, the company developed the prototype for what would eventually be Houseparty—an app that started broadcasting using the device's front camera as soon as it was opened, while also notifying friends that you were live.

Pivot Validation

Since Houseparty was targeting a younger demographic, Sima Sistani—the company's new COO—led the team in visiting as many schools as possible to get feedback on the app. They had an engineering goal to tweak the app and have a new version ready at the end of every week. For example, the feature to 'lock' a room for privacy was added during this process. The beta version was also rolled out to close family and friends of the core team—all of whom had very positive reactions.

Commit

Length of Pivot

Houseparty was actually launched in stealth on Android and iOS with the developer listed as "Alexander Herzick" (quick trivia: this was actually the COO's husband who was chosen because of his minimal online existence). Just around six months after it was launched, it grew to 1 million DAU. The numbers also showed that the frequency of usage was high—users were logging into the app several times a week.

Pivot Funding

Around six months after the pivot, in late 2016, Sequoia led a $50 million round for the company. Existing investors Greylock Partners, Aleph, and Comcast Ventures also participated.

Pivot Outcomes

Houseparty eventually plateaued because it wasn't growing fast enough to monetize users or raise more venture dollars. Around 3 and a half years after the pivot, Fortnite-creators Epic Games acquired Houseparty for around $35 million. Soon after, Epic Games released a Houseparty feature into Fortnite to allow gamers to see livestreams of their friends playing. When the pandemic hit, Houseparty's popularity skyrocketed—gaining 50 million new registrations in May 2020. Once the hype died down, people migrated to different apps (Zoom for work, Clubhouse for social interactions, etc.) and its numbers were on the decline again. In October 2021, Epic Games discontinued the app and is reportedly shifting gears to creating social experiences in the metaverse.

Lessons Learned

1. **Growth without retention is dangerous.** When Rubin made the decision to pivot, Meerkat had recently raised

around $14 million and was still growing at a rate of around 20-30%. But he was looking for weaknesses. He noticed that the stats of regular users broadcasting had nosedived. He also realized that the livestream-competitors launched by Twitter and Facebook would quickly cannibalize the celebrities and influencers on Meerkat. Even though they could have continued operations for at least a few more months, he decided to deal with reality head on. If you cannot get an average person to use your product on a daily basis, the company probably will not work. It's important to recognise the early signs of failure and take proactive steps immediately.

2. **Friends and family are an invaluable sample. Listen to them.** The team rolled out an early version of Houseparty to close family and friends to get direct feedback on it. This go-to-market strategy was very different from the hype that surrounded Meerkat's launch. This is the best kind of product development because it reduces the number of pitfalls from the start.

Discord (Fates Forever)

Mobile gaming → VoIP + Messaging

Original premise

As the popularity of the iPad grew, Jason Citron saw that there were no core video games developed to be played on tablets. Most of the games at the time were designed for desktop PCs. Predicting that the next generation of gamers would grow up using tablets, he saw this as an opportunity to build a gaming company around the idea of tablets and core multiplayer games. While developing their first game, Fates Forever, the team also built in basic communication tools (voice and text chat) so users could talk to each other while they played.

Align

Warning Signs

Although the game Citron's team created got great reviews, tablet gaming didn't catch on. Gamers just weren't looking to play games on their iPads. This meant that the company wasn't seeing the traction they needed. Their performance metrics also didn't look good.

Time to Pivot

They started thinking about pivoting around six months after they launched.

Initial Funding / Team

Pre-pivot Citron had raised $1.1 million of seed money and about $13 million more in their Series A and B rounds. They had a team of around 10 people.

Explore

New Opportunity

Even though interest for the game itself flagged, Citron and his team realized that a lot of users liked the inbuilt text and voice features. Most gamers at this point were using Skype or TeamSpeak to communicate—and they all hated it. Neither of those applications provided consistent low-latency and high-security communication. The team was sure they could build something better and ended up with Discord.

Pivot Validation

Immediately after it launched, Discord had only around 10 users on an average day. Then someone posted about it in the Final Fantasy XIV subreddit with a link to a Discord server where they could talk about features. Seeing an opportunity, Citron and Stan Vishnevskiy joined the server and started talking to the redditors who showed up. This drove up more than 500 registrations. Doubling down on the interest, just a week later, Discord also hosted an AMA with early users to take requests for features.

Commit

Length of Pivot

Around six months after the AMA, Discord had a whopping 3 million registered users. It also recorded growth of 1 million new users joining each month. Fast forward to its first anniversary and 11 million users were using Discord to send around 40 million messages per day on the platform.

8. Pivot Funding

After pivoting, Discord raised money for the first time in January 2016—a $20 million Series C round led by Greylock and Spark Capital at a valuation of $100 million.

Pivot Outcomes

After the pivot, Discord slowly started exploring ways to monetize its users. Its primary source of revenue is "Nitro," a paid subscription service through which users can enjoy perks like customizing their avatars, using emojis, and better video resolution. Interestingly, long term members of the Discord community have stated that although this tier doesn't add much tangible value, they stay subscribed to support the company. In 2018 Discord also experimented with curating and selling games to users. This strategy didn't work and was shut down in just a few months.

As the platform grew, people inevitably also used it for unsavory purposes. The Discord team admitted to catching on to this late, and like most social networks continue to fight the uphill battle against extremist hate groups.

When the pandemic hit, the number of users on Discord skyrocketed—there was a 47% increase from February to July 2020. With most of the world in quarantine, the company's use case expanded dramatically as IRL communities migrated online. Discord saw book clubs, study groups, and sometimes people just hanging out with a group of their friends. Taking cue from the increased non-gamer user base, the company rebranded itself. Its new tagline—"Your place to talk."

At the beginning of this year, the company was valued at around $15 billion. Forbes also listed it as an IPO to watch out for in 2022.

Lessons Learned

1. **Don't wait too long to have the tough conversations.**
   When Citron realized that the video game they had designed
   wasn't doing well, he was very transparent with his investors.
   He told them about the flagging numbers and engagement
   even before deciding to pivot. They ended up helping him
   through the process of what to do next. These are difficult
   conversations to have, but it's crucial to address the problem
   head on as soon as possible.

2. **Nail that one thing you're banking on.** The early version
   of Discord actually lacked some ancillary features that their
   competitors had. But its main offering — voice chat — was
   absolutely spot on. The app's audio quality and real-time
   sync was miles ahead of its time. After launching, Citron's
   team began releasing smaller miscellaneous features almost
   daily. When you release your product, make sure you build
   out that one feature really well. The rest can and will fall into
   place later.

ACKNOWLEDGEMENTS

Thank you to Ben Putano, Cedric Chin, Elisa Mala, Erica Schneider, Herbert Lui, Kasey Jones, Mariko Gordon, Nir Eyal, Rachel Jepsen, and Visakan Veerasmy for inspiring me as a writer.

Thank you to beta readers Charlene Wang, George Chewning, Johnson Lin, Kamil Nicieja, Kurt Thigpen, Miroo Kim, Randall Bennett, Sanketh Andhavarapu, Wayne Gerard, Zeph Chang for your comments and feedback.

To Rhea Purohit, your research, thinking, and editorial support have been an integral part of this book. The power of a good cold DM!

To my clients past and present, thank you for trusting me to hold your stories and dreams. Your ambition and compassion fire me up.

To my parents Anping Shen and Shixin Mao, thank you for your love and support and for having the courage to pivot from a life in China so that I could thrive. And my sister Amy for always keeping it real.

To my wife Amanda Phingbodhipakkiya, thank you for being my partner in all things and showing me everyday what is possible with courage, imagination, and a full heart. I love you so much and am so proud to be on your team. **Jason**

Thank you Jason, for believing in me, for showing me how to operate with kindness and enthusiasm, and most of all, for always pushing me beyond my comfort zone, which I believe is how we grow.

And to my partner Yash, thank you for creating a space for me to become a better version of myself. Real love and support is incredibly empowering, and you exemplify that. **Rhea**

JASON SHEN is a writer, speaker, and executive coach. As CEO of Refactor Labs, he works with outlier founders and executives to lead through hard pivots in their business and career, so they can make a dent in the universe.

Jason's bylines have appeared in *Every*, *Vox*, *CNBC*, *TechCrunch*, and *Fast Company*. He's spoken at Pinterest, GrubHub, Procter & Gamble, and Cornell University. A three-time founder, Jason's last startup, Midgame, an AI gaming tools company, was acquired by Meta in 2020.

Jason was a national champion gymnast at Stanford University where he earned a BS and MS in biology. A lifelong athlete, he's broken two Guinness world records and was recognized as one of the fittest people in the tech industry by *Men's Health* and *Outside Magazine*. He lives in Brooklyn with his wife Amanda, a transdisciplinary artist.

Learn more at jasonshen.com and pathtopivot.com.